Metric Edition

BASIC MATHEMATICS

Book Three

A. L. GRIFFITHS

Illustrated by G. B. Hamilton

OLIVER AND BOYD LTD
EDINBURGH

First published 1968
Metric Edition 1969

© *A. L. Griffiths 1969*

05 002142 7

MADE AND PRINTED IN GREAT BRITAIN BY
THOMAS NELSON (PRINTERS) LTD.,
LONDON AND EDINBURGH

CONTENTS

CONTENTS

WHAT IS A UNIT?

Here is one button: ☺ . The Latin word for ' one ' is *Unus*. From this Latin word we get our own word *unit*, which also means one.

Here are 3 ones or 3 units: ○○○. Here are 4 units: △△△. Here are 9 units: ○○ ○○ ○○○○○

WHAT IS A TEN?

Here are some units: ◇ □ ◇□◇ □□ ◇ ◇ □ ◇. Count them and you will find there

are 10. If these units are arranged in a line we can see more clearly that they make

1 ten ⬛⬛⬛⬛⬛⬛⬛⬛⬛⬛.
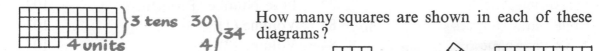

$$1 \times 10 = 10 \qquad 10 \times 1 = 10$$

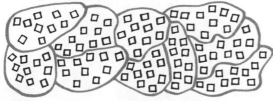

}3 tens   30}
}34
4 units   4}

How many squares are shown in each of these diagrams?

(a)   (b)   (c)

WHAT IS A HUNDRED ?

Here are a hundred squares.
It was very difficult to count them
and group them in tens.

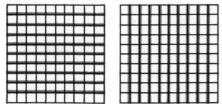

When the squares are arranged in rows of
tens we can see clearly that 10 tens make 1
hundred.

$$10 \times 10 = 100$$

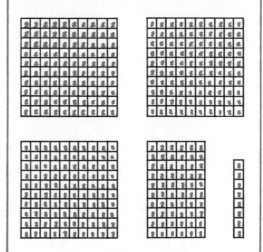

How many stamps can you see
in this diagram ?

Write the number in numerals.

WHAT IS A THOUSAND ?

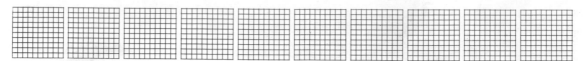

Look at the small squares above. There are a hundred small squares in each large square and so we can count them in hundreds, like this : 1 hundred, 2 hundreds, 3 hundreds, 4 hundreds, 5 hundreds, 6 hundreds, 7 hundreds, 8 hundreds, 9 hundreds, 10 hundreds.

Ten hundreds is written in figures like this : 1000. Ten hundreds is the same as one thousand.

$$10 \times 100 = 1000$$

A   Use the diagram to answer these questions.

    1. How many less than a thousand is (*a*) 990 (*b*) 899 (*c*) 909 ?
    2. What number is half of a thousand ?
    3. What number is a fifth of a thousand ?
    4. What number is a tenth of a thousand ?
    5. What number must be added to 320 to make a thousand ?
    6. What number must be taken from a thousand to leave 460 ?
    7. From a thousand take 330 and 370.

WHAT IS TEN THOUSAND ?

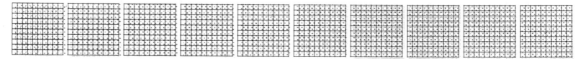

Imagine there are 10 dots in each of the small squares.
In each large square there are $100 \times 10 = 1000$ dots.
In the ten large squares the number of dots is : $1000 \times 10 = 10\ 000$

B   Use the diagrams on this page to answer these questions :

    1. What number must be added to each number in Row A to make a thousand ?
    2. What number must be added to each number in B to make ten thousand ?

Row A	900	300	750	999	890	99
Row B	9900	9090	9999	9001	8500	5000

WHAT IS A HUNDRED THOUSAND ?

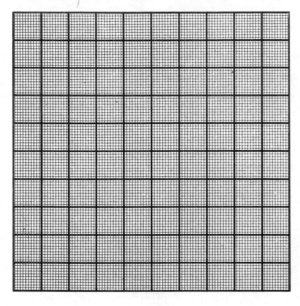

Here are ten thousand small squares. If ten dots were placed in each of the small squares the total number of dots would be :

$$10\,000 \times 10 = 100\,000$$

A HUNDRED THOUSAND

How many dots would there be in this square?

How many dots would there be in this square?

WHAT IS A MILLION?

If a hundred very tiny dots could be placed in each of the small squares in the diagram above, the total number of dots would be :

$$10\,000 \times 100 = 1\,000\,000$$

$$100\,000 \times 10 = 1\,000\,000$$

$$1000 \times 1000 = 1\,000\,000$$

This artificial satellite has travelled a MILLION kilometres in space.

The distance from London to Copenhagen by air is nearly a thousand kilometres. An aircraft which has travelled between these two capitals a thousand times will have flown nearly a MILLION kilometres.

COPENHAGEN

1000 km

LONDON

MILLIONS	HUNDRED THOUSANDS	TEN THOUSANDS	THOUSANDS	HUNDREDS	TENS	UNITS
1	0	0	0	0	0	0
	1	0	0	0	0	0
		1	0	0	0	0
			1	0	0	0
				1	0	0
					1	0
						1

A   Add these numbers without copying them down :

     1. 4000 + 700 + 47   *3.* 17 000 + 600 + 50 + 8

     2. 30 000 + 5000 + 90 + 6   *4.* 90 000 + 300 + 60 + 2

B   Write in words the difference between these pairs of numbers :

     1. 483 672 and 463 672   *2.* 1 207 334 and 1 907 334   *3.* 72 631 and 74 731

C   Write in numerals the number that must be added to

     1. 5634 to make 6644 ;   *2.* 80 340 to make 90 350 ;

     3. 304 471 to make 505 471.

D   Here are the kilometre readings of some cars at the beginning and end of a month. Find how many miles each has travelled during that time. Write the answers only.

Car 1	3479	4499
Car 2	17603	20703
Car 3	71294	71996
Car 4	111725	112825

ROUNDING OFF NUMBERS

We have already learned that we can round off a number like 7835 to the nearest thousand. We would call it 8000, because 7835 is nearer to 8000 than it is to 7000. The number 7435 when rounded off to the nearest thousand is 7000, because it is nearer 7000 than 8000.

E   Here are the approximate areas of some countries in square kilometres. Round off these numbers to the nearest million.

     Australia 7 616 000   *Brazil* 8 419 000   *Canada* 9 846 000   *U.S.A.* 9 254 000

F   Here are some astronauts' speeds. Round them off to the nearest thousand.

     Titov 27 766 km/h   *Glen* 28 064 km/h   *Gagarin* 27 837 km/h

G   Here are the lengths of some famous bridges in metres. Round off the numbers to the nearest hundred.

     Menai 173   *London* 306   *Severn* 987   *Sydney Harbour* 503

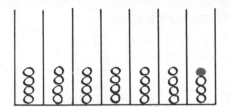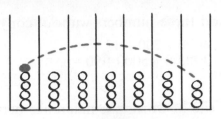

In the first abacus picture you can see there are four counters in each column. In the second abacus one counter has been taken from the units column and placed in the millions column.

So we have added a million and taken away one : $1\,000\,000 - 1 = 999\,999$.

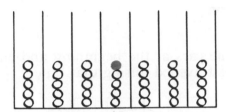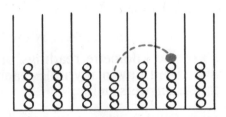

These abacus pictures show that one counter has been removed from the thousands column and placed in the tens column. We have subtracted 1000 and added 10. This means we have subtracted 990.

A Draw abacus pictures to show that by moving a counter from one column to another you can :

1. add 999 990   *2.* add 99 990   *3.* add 999 000   *4.* add 999 900

5. subtract 9900   *6.* subtract 99 000   *7.* subtract 9999   *8.* subtract 90 000

B Now try these more difficult examples :

1. Add 19 998 by moving 2 counters from one column to another.

2. Add 4950 by moving 5 counters from one column to another.

3. Subtract 2997 by moving 3 counters from one column to another.

4. Subtract 399 996 by moving 4 counters from one column to another.

Make a work card for your friends by thinking of some examples of your own. A good way to experiment is to draw an abacus on a sheet of paper and use counters to move from column to column.

A   Study the abacus numbers in the diagram.

We can see that the value of the numeral 6 depends on its position.

Now look at the numerals in these numbers :   8763

                      3678

You can see that the same numerals have been used in each number, but they are in different positions.

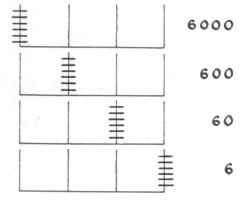

1. How many times greater is the 8 in the top number than the 8 in the bottom number ?

2. How many times as big as the 7 in the bottom number is the 7 in the top number ?

3. How many times greater is the 3 in the bottom number than the 3 in the top number ?

4. How many times as big as the 6 in the top number is the 6 in the bottom number ?

5. What are the greatest and the lowest numbers you can make with the numerals 7, 6, 5 and 9 ?

B   Here are two more numbers shown on the abacus. We know that the 5 in the top number is ten times as big as the 5 in the bottom number. Or we can say that the 5 in the bottom number is $\frac{1}{10}$ as big as the 5 in the top number.

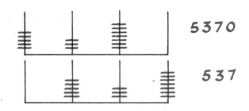

The 3 in the bottom number is $\frac{1}{10}$ the value of the 3 in the top number.
The 7 in the bottom number is $\frac{1}{10}$ the value of the 7 in the top number.

1. Multiply these numbers by 10 : 7,   11,   23,   493,   240,   5000.

2. Multiply these numbers by 100 : 3,   20,   37,   4009,   130.

3. Find $\frac{1}{10}$ of 270,   7300,   90,   4350,   600,   780.

4. Find $\frac{1}{100}$ of 7300,   800,   6000,   14 000,   10 000,   11 100.

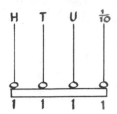

What will the next place to the right of the units place show ? It will show tenths of a unit.

On this abacus there is a dotted line to separate the whole numbers from the fractions.

When we write down a number we place a 'dot' between the units and tenths : 111·1. This dot is called the DECIMAL POINT.

213·5 is read 'two hundred and thirteen point five'. It can also be written $213\frac{5}{10}$. The decimal point is printed higher than a full stop.

> Digits to the left of the decimal point are whole numbers.
> Digits to the right of the decimal point are fractions.

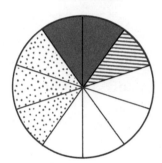

This circle has been divided into ten equal parts. Each part is one tenth, $\frac{1}{10}$, of the whole circle.

$\frac{1}{10}$ written as a decimal is 0·1.

We can see that $\frac{2}{10}$ or 0·2 of this circle has been coloured.

1. What decimal fraction of the circle has been :
   (a) left blank ? (b) covered with dots ?
   (c) shaded with horizontal lines ?

2. What decimal fraction of each of these drawings is (a) shaded (b) unshaded ?

  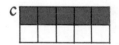 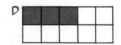

3. Look at these drawings below.

We can see in the first drawing there are 3 whole circles and $\frac{3}{10}$ which are shaded. This can be written $3\frac{3}{10}$, or in decimal form, 3·3.

Look at the other drawings.

Write down in decimal form the circles and parts of circles that are shaded.

(a)

(b)

(c)

(d)

(e)

A Tim bought a new trip indicator for his bicycle and decided to check its accuracy by cycling from one kilometre mark to the next. The diagram shows the indicator reading at the start. As he rode along, Tim realised that the coloured figures showed tenths of a kilometre. Just as he reached the end of the measured kilometre, the figure 9 in the tenths column changed to 0 and 1 appeared in the units column. This meant 1 whole kilometre, or 10 tenths ($\frac{10}{10}$). Tim returned along the same route. Midway between the kilometre marks he noticed his indicator reading was 0015. This means $\frac{15}{10}$ or $1\frac{5}{10}$ or $1\frac{1}{2}$ or 1·5 kilometres. What was the reading on the indicator when he returned to the starting point?

0	0	0	0
0	0	0	1
0	0	0	2
0	0	0	3
0	0	0	4
0	0	0	5
0	0	0	6
0	0	0	7
0	0	0	8
0	0	0	9
0	0	1	0

B Now see if you can write these in decimal form :

1. $\frac{11}{10}$ *2.* $4\frac{7}{10}$ *3.* $\frac{23}{10}$ *4.* $\frac{31}{10}$ *5.* $\frac{74}{10}$
6. $12\frac{7}{10}$ *7.* Eight and nine-tenths   *8.* Twenty and five-tenths

If you have a hand calculating machine, use this to show decimal fractions.

C Make sure your ruler measures centimetres and tenths of a centimetre. Remember, one tenth can be written as a vulgar fraction $\frac{1}{10}$; or as a decimal fraction, 0·1.

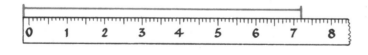

You can see that this line is $7\frac{2}{10}$ centimetres long. In decimal form we would write 7·2 centimetres (7·2 cm).

Measure these lines carefully and write down their lengths in decimal form :

1. ——————————————————————

2. ————————————————

3. ————————————————————————

4. ————————

5. ————————————————————

Write these abacus numbers :

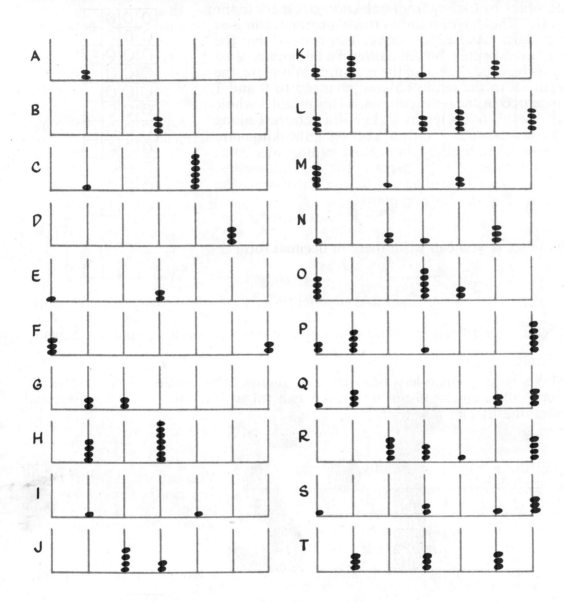

Make rough abacus drawings to show these numbers :

1. One million, four hundred and four thousand
2. Seven hundred thousand and seventeen
3. Two million, two hundred and twenty
4. Three hundred and three thousand, and three
5. Eighty thousand and eighty

6. 7 039 000
7. 2 200 404
8. 705 069
9. 30 003
10. 1 010 100

TENS	UNITS	TENTHS
O		O
O	O	
	O	O
OO		
	OO	
		OO

A
B
C
D
E
F

1. Here is a simple decimal abacus. It shows how many numbers can be made by using 2 counters at a time.
Write these numbers in numerals like this :

        A 10·1

2. Draw a large abacus on a separate sheet of paper and see how many numbers you can make by using 3 counters at a time.

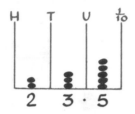

The first bead abacus shows the number 23·5. The second abacus shows the number 235.

    23·5 × 10 = 235
    235 ÷ 10 = 23·5

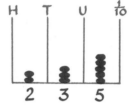

3. Multiply these numbers by ten :
    0·3,   1·1,   0·9,   1·3,   14,   27·1,   10·2,   301·8

4. Divide these numbers by ten :
    23,   407,   6,   66,   101,   2012,   1001

5. Study these numbers carefully. Arrange them in order of size, starting with the smallest :
    (*a*) 10·9,   1·9,   90·1,   9·1,   109,   901
    (*b*) 33,   33·3,   333,   3·3,   30·3
    (*c*) 707,   77,   7·7,   70·7,   77·7
    (*d*) 90·9,   90,   9·0,   0·90,   900

6. What must be added to each of these numbers to make it up to 1 ?
    0·3,   0·9,   0·5,   0·8,   0·6

7. What must be added to each of these numbers to make it up to 10 ?
    1·6,   3·3,   7·5,   0·7,   9·9,   1·1

8. What must be added to each of these numbers to make it up to 100 ?
    78·9,   43·1,   10·1,   50·5,   5·5,   0·9

A   Write these speedometer readings in words. Remember that the coloured figures represent tenths.

1. 0 0 2 1 0 6   *2.* 0 3 4 0 6 7   *3.* 0 1 8 0 0 1   *4.* 0 0 8 7 6 4

B   Write these abacus numbers in decimal form, using numerals :

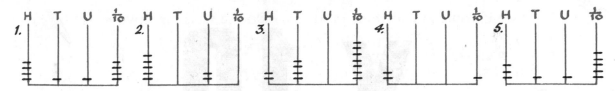

C   Study these cyclometer readings carefully. Think what each of the readings will be after the next $\frac{1}{10}$ kilometre and write it in your book.

1. 0 9 9 9   *2.* 3 9 9 0   *3.* 0 1 0 9   *4.* 1 0 0 0   *5.* 1 9 9 9

D   Look at these numbers and write down the number which should come next in each line.

1. 0·2   0·4   0·6   0·8   □   *3.* 0·5   1   1·5   2   □
2. 0·3   0·6   0·9   1·2   □   *4.* 2·1   4·2   8·4   □

E   *1.* Here are the recorded times, in seconds, for the first four runners in a 100-metres race. Arrange the times in order, starting with the winner.

11·4,   11·0,   11·5,   10·8

2. Until recently Jeff's best time for the 100-metres' sprint was 13·1 seconds. He has now clipped 0·7 of a second off this time. What is his new record time?

3. A space-rocket travelled 7 kilometres in one second. How far did it travel in 0·1 of a second?

F

BLUE LAKE
2·3 kilometres
5·4 kilometres
BLACK PARK
6·2 kilometres
BEECH WOOD

Bob cycled from Blue Lake to Beech Wood, from Beech Wood to Black Park, and then back to Blue Lake.
1. How many kilometres had he cycled?
2. How much farther is it from Blue Lake to Beech Wood than from Black Park to Blue Lake?
3. How much farther is it from Beech Wood to Black Park than from Blue Lake to Beech Wood?

A   When we multiply by ten we move the digits 1 place to the left.

H	T	U
	1	7

H	T	U
1	7	

$17 \times 10 = 170$

To multiply a number by twenty, we move the digits one place to the left and multiply by 2 : $17 \times 20 = 340$

Now see if you can find these products.   Do the working in your head.

1. 47×10 2. 10×16 3. 50×10 4. 5463×10
5. 10×300 6. 23×20 7. 20×32 8. 33×30
9. 3314×20 10. 14×40 11. 15×50 12. 60×13

B   To multiply   23 by 13
we can multiply 23 by 10, which gives 230,
then multiply   23 by 3, which gives   69,
and find the total, which gives   299.

We usually set   23
down the working $\times 13$
like this :   230
                   69
                299

Here is another   39
example : $\times 23$
                780
                117
                897

Work these examples :

1.   23   2.   47   3.   26   4.   63   5.   29
   $\times 15$ $\times 19$ $\times 13$ $\times 18$ $\times 16$

6.   79   7.   65   8.   76   9.   42   10.   65
   $\times 28$ $\times 43$ $\times 84$ $\times 37$ $\times 65$

C   Now try these examples with 3 digits in the multiplicand :

1.   469   2.   378   3.   485   4.   986   5.   949
   $\times 26$ $\times 43$ $\times 17$ $\times 58$ $\times 29$

6.   706   7.   980   8.   409   9.   724   10.   716
   $\times 43$ $\times 67$ $\times 79$ $\times 86$ $\times 94$

·How do we multiply a whole number by a hundred ? We move the digits two places to the left and put a zero in the units place and tens place of the answer.

$$\begin{array}{r} 14 \\ \times 100 \\ \hline 1400 \end{array} \qquad \begin{array}{r} 14 \\ \times 200 \\ \hline 2800 \end{array} \qquad \begin{array}{r} 17 \\ \times 500 \\ \hline 8500 \end{array}$$

A   Finish these in the same way :

    1. $7 \times 100 = ?00$ *2.* $9 \times 200 = ?00$
    3. $28 \times 100 = ?00$ *4.* $16 \times 200 = ?00$

B   Work these examples in your book :

  1. 47×200 *2.* 300×86 *3.* 400×23 *4.* 700×82
  5. 631×100 *6.* 300×236 *7.* 400×178 *8.* 132×600

C

$$\begin{array}{r} 476 \\ \times 192 \\ \hline \end{array}$$

$476 \times 100 = \underline{47\ 600}$ { We show the product is hundreds by putting a zero in the tens place and units place.
$476 \times 90 = 42\ 840$ { We show the product is tens by putting a zero in the units place.
$476 \times 2 = \underline{952}$
$476 \times 192 = \underline{\underline{91\ 392}}$

Now try these examples :

  1. $\begin{array}{r} 213 \\ \times 312 \\ \hline \end{array}$ *2.* $\begin{array}{r} 476 \\ \times 243 \\ \hline \end{array}$ *3.* $\begin{array}{r} 789 \\ \times 654 \\ \hline \end{array}$ *4.* $\begin{array}{r} 956 \\ \times 386 \\ \hline \end{array}$ *5.* $\begin{array}{r} 477 \\ \times 639 \\ \hline \end{array}$

D   Here are some examples with zeros in the multiplier. If we really understand what we are doing the zeros make the working easier still.

$$\begin{array}{r} 476 \\ \times 300 \\ \hline 142\ 800 \end{array} \qquad \begin{array}{r} 476 \\ \times 306 \\ \hline 142\ 800 \\ 2\ 856 \\ \hline 145\ 656 \end{array} \qquad \begin{array}{r} 476 \\ \times 360 \\ \hline 142\ 800 \\ 28\ 560 \\ \hline 171\ 360 \end{array}$$

Now work these examples. Check each answer by changing the places of the multiplier and multiplicand :

  1. $\begin{array}{r} 839 \\ \times 460 \\ \hline \end{array}$ *2.* $\begin{array}{r} 765 \\ \times 830 \\ \hline \end{array}$ *3.* $\begin{array}{r} 749 \\ \times 900 \\ \hline \end{array}$ *4.* $\begin{array}{r} 635 \\ \times 906 \\ \hline \end{array}$ *5.* $\begin{array}{r} 787 \\ \times 307 \\ \hline \end{array}$

A   Write out these sentences, putting in the largest whole number which can replace □. Do it like this : $5 \times 6 < 33$.

     1. $12 \times □ < 90$ *2.* $□ \times 8 < 39$ *3.* $□ \times 6 < 53$ *4.* $□ \times 7 < 83$

     5. $11 \times □ < 65$ *6.* $12 \times □ < 100$ *7.* $9 \times □ < 96$ *8.* $7 \times □ < 60$

B   We want to find the quotient in the division example $70)\overline{225}$.. We must find the greatest whole number that can replace □ in : $70 \times □ < 225$.

$$\begin{array}{r} 3 \\ 70)\overline{225} \\ 210 \\ \hline 15 \end{array}$$

Here are 3 seventies,
When 3 seventies are subtracted,
the remainder is 15.

We can check our answer by multiplying.

$$\begin{array}{r} 70 \\ \times 3 \\ \hline 210 \\ 15 \\ \hline 225 \end{array}$$

Our calculating machine would solve this problem by subtracting 70 three times.

2⟮2⟮5⟮
1⟮5⟮5⟮ −(1 × 70)
⟮8⟮5⟮ −(2 × 70)
⟮1⟮5⟮ −(3 × 70)

Write out these number sentences, putting in the largest whole number which can replace □ :

     1. $60 \times □ < 367$ *2.* $30 \times □ < 200$ *3.* $70 \times □ < 340$

     4. $90 \times □ < 850$ *5.* $40 \times □ < 310$ *6.* $50 \times □ < 290$

     7. $80 \times □ < 950$ *8.* $80 \times □ < 700$ *9.* $20 \times □ < 230$

     10. $60 \times □ < 630$ *11.* $90 \times □ < 486$ *12.* $70 \times □ < 833$

C   Now work out the following examples.

To find the quotient, we must first find the largest whole number that will replace □ : e.g. $60 \times □ < 135$.

     1. $60)\overline{135}$ *2.* $70)\overline{568}$ *3.* $40)\overline{283}$ *4.* $30)\overline{267}$ *5.* $80)\overline{378}$

D   Work out these division examples as a calculating machine would do them.

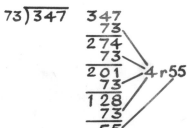

     1. $67)\overline{273}$ *2.* $45)\overline{388}$ *3.* $29)\overline{176}$

     4. $47)\overline{370}$ *5.* $26)\overline{215}$ *6.* $39)\overline{197}$

     7. $63)\overline{258}$ *8.* $56)\overline{319}$ *9.* $48)\overline{277}$

     10. $49)\overline{311}$ *11.* $59)\overline{278}$ *12.* $98)\overline{219}$

ESTIMATING THE QUOTIENT FIGURE

In the example $39\overline{)93}$, we can see that the divisor 39 is closer to 40 than it is to 30. If we call it 40, the quotient figure is much easier to work out, for we know that 2 forties make 80.

A   See if you can work out the quotient figure correctly first time in these examples :

1. $37\overline{)83}$ 2. $28\overline{)90}$ 3. $18\overline{)59}$ 4. $27\overline{)91}$ 5. $38\overline{)90}$

6. $41\overline{)87}$ 7. $32\overline{)95}$ 8. $19\overline{)87}$ 9. $37\overline{)80}$ 10. $28\overline{)93}$

B   Divide and then check your answers by multiplying :

1. $48\overline{)163}$ 2. $39\overline{)187}$ 3. $38\overline{)261}$ 4. $19\overline{)173}$ 5. $69\overline{)224}$

6. $28\overline{)247}$ 7. $47\overline{)253}$ 8. $37\overline{)371}$ 9. $18\overline{)169}$ 10. $48\overline{)463}$

11. $98\overline{)700}$ 12. $87\overline{)466}$ 13. $77\overline{)570}$ 14. $66\overline{)643}$ 15. $57\overline{)437}$

C   Now for some bigger quotients :

$$\begin{array}{r} 35 \\ 28\overline{)980} \\ 84 \\ \hline 140 \\ 140 \\ \hline \end{array}$$

1. $22\overline{)682}$ 2. $33\overline{)693}$ 3. $41\overline{)492}$ 4. $15\overline{)630}$

5. $25\overline{)375}$ 6. $70\overline{)560}$ 7. $23\overline{)368}$ 8. $33\overline{)759}$

9. $34\overline{)918}$ 10. $54\overline{)702}$ 11. $14\overline{)616}$ 12. $16\overline{)512}$

D   Now for some bigger quotients with remainders :

$$\begin{array}{r} 23\,r1 \\ 36\overline{)829} \\ 72 \\ \hline 109 \\ 108 \\ \hline 1 \end{array}$$

1. $21\overline{)530}$ 2. $34\overline{)785}$ 3. $30\overline{)466}$ 4. $23\overline{)369}$

5. $29\overline{)686}$ 6. $41\overline{)699}$ 7. $29\overline{)960}$ 8. $76\overline{)989}$

9. $35\overline{)843}$ 10. $19\overline{)877}$ 11. $72\overline{)940}$ 12. $28\overline{)673}$

E   In these examples the quotients are given ; can you find the divisors ?

1. $\triangle\overline{)192}$ quotient 8   2. $\triangle\overline{)147}$ quotient 21   3. $\triangle\overline{)688}$ quotient 43   4. $\triangle\overline{)429}$ quotient 33   5. $\triangle\overline{)323}$ quotient 19

F   Here are some examples with the same number for the quotient as for the divisor. Can you find each number ?

1. $\square\overline{)100}$ quotient $\square$ 2. $\square\overline{)64}$ quotient $\square$ 3. $\square\overline{)400}$ quotient $\square$ 4. $\square\overline{)144}$ quotient $\square$ 5. $\square\overline{)169}$ quotient $\square$

Some boys decided to collect 20 conkers for their nature table. This pictogram shows the number each boy collected.

TIM   ● ● ●
BOB   ● ● ● ●
PETER   ● ● ● ● ● ●
DAVID   ● ● ● ● ● ●

If the boys had each collected the same number, Tim would have collected 2 more and David and Peter would each have collected one less. We can see that each boy would have collected 5 conkers.

TIM   ● ● ● ● ●
BOB   ● ● ● ● ●
PETER   ● ● ● ● ●
DAVID   ● ● ● ● ●

We say that the boys collected an AVERAGE of 5 conkers each.

o o o o o o o o o o o o
o o o o o o
o o o o o o
o o o o

Here are some counters. You can see that there are 12 in the top line, 6 in the second, 6 in the third and 4 in the bottom line.

o o o o o o o
o o o o o o o
o o o o o o o
o o o o o o o

Now the counters have been moved about until each line has the same number.
Can you see that the *average* of 12, 6, 6 and 4 is 7 ?

A   Take a box of counters and see if you can find the average of these numbers in the same way :

1. 7, 7 and 10   *2.* 6, 8, 9 and 1   *3.* 8, 4 and 3

B

Look at these rods carefully. You can see that the top line is made up of a 5 rod, a 4 rod, a 3 rod, a 7 rod and a unit rod.
The line of rods is 20 units long.
The bottom line is also 20 units long and is also made up of 5 rods, but this time they are each the same length.
What is the average of 5, 4, 3, 7 and 1 ?
If we use numbers instead of rods or counters, we can find the average just as easily.

In this example we must find one number that can replace each of the three numbers and give the same sum.

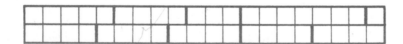

$$\overset{5}{\cancel{2}} + \overset{5}{\cancel{8}} + \overset{5}{\cancel{4}} = 15$$

A   Find the average of the numbers in the sets below, like this :

$$7, 9, 2 \ (\ \overset{6}{\cancel{7}} + \overset{6}{\cancel{9}} + \overset{6}{\cancel{2}} = 18 \ . \quad \text{The average is } 6)$$

(*a*) 6, 8, 10   (*b*) 9, 11, 1   (*c*) 5, 5, 7, 3

B   Ann's marks in a dancing examination were :

BALLET	5	CLASSICAL	9	MODERN	10

Ann had a total of 24 marks. If she had gained the same number of marks in each dance, she would have had $(24 \div 3)$ marks, that is 8 marks, for each dance. Ann's average mark for the 3 dances was 8.

Here are four 'hands' of bananas. Give the average number of bananas in a hand.

C

You will need squared paper.

1. Draw a pictogram to show the heights in centimetres of 6 people in your class. Find the average height of these 6 pupils.

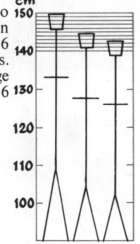

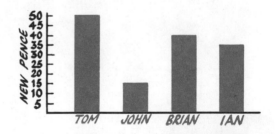

2. Draw a bar graph to show the pocket-money received by 8 people in your class. Work out the average amount of pocket-money.

The Romans used these seven letters as symbols to write numbers :

I	V	X	L	C	D	M
1	5	10	50	100	500	1000

If you study the following rules they will help you to read and write numbers in the Roman system.

Numerals of the same value are added :
    XXX (30)   MM (2000)   CCC (300)

The same symbol is rarely used more than 3 times together.
IIII is mostly written IV.

A numeral placed after one of greater value is always added :
    XI (11)   XXI (21)   MC (1100)   VII (7)

A numeral placed before one of greater value is always subtracted :
    IV (4)   XL (40)   XC (90)

When a numeral is placed between two numerals of greater value, it is always subtracted from the numeral immediately following it.
    XXIX (29)   CIV (104)

It is easy to write numbers in Roman numerals by combining the symbols in order. We can write 1993 by putting together M (1000), CM (900), XC (90), and III (3):
    MCMXCIII

A   Write the numbers of these volumes in Arabic numerals.

B   Write the dates on these cornerstones in Arabic numerals.

C   Here is a number line with Roman numerals. Copy the number line and write in the missing numerals.

D   Here is another number line. Copy the line and write in the missing Roman numerals.

The Romans found difficulty in calculating with their numerals. They usually used the abacus for this purpose. They wrote only the answer in numerals. The Latin word for pebble is 'calculus', for it was with pebbles and an abacus that the Romans did their calculating.

The abacus used by the Romans was sometimes a wooden board on which grooves had been cut to take the rounded pebbles. Sometimes it was a board with lines on which could be placed thick discs, like the pieces used in a game of draughts.

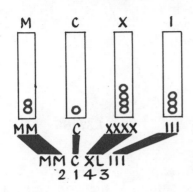 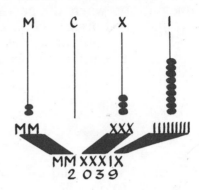

A Write these abacus numbers in Roman numerals :

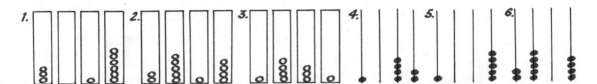

B Addition and subtraction are quite easy with Roman numerals. The Romans would be able to do calculations like these without the abacus :

Addition				*Subtraction*			
DC	LXXX	VIII		M	DCC	L	II
C	XX	VII			C	X	VI
DCCC	X	V		M	DC	XXX	VI

Multiplication and division were too difficult to work in Roman numerals. For these calculations the abacus was always used.

Work out these examples in Roman numerals. Then write out the examples in Arabic numerals and check your answers.

1. Add DCC XC VII
          DCC LX VI

2. Subtract DCCC LXX IV
                 C LXXX V

> You will need thread, a washer, various weights and a stop watch.

Make a pendulum by tying a weight to the end of a length of thread. This weight is called a ' bob '. You can use a washer, an iron nut or even a ball of plasticine. Fasten the end of the thread to a hook and place the hook in a position where the pendulum is free to swing to and fro. A good place to choose is an open doorway. The length of the pendulum is measured from the point of support to the centre of the weight.

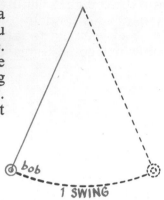

1. Tie a washer to the end of a metre length of thread. Time 20 swings with a stop watch. (You will need a partner to do the timing while you call out the number of swings.) Do not have larger swings than shown in the drawing.
Tie two washers to the length of thread and time 20 swings.
Tie three washers to the length of thread and time 20 swings.
Does the weight make any difference to the time of the swings?

2. Tie any weight to a 120 cm length of thread. Time 20 swings.
Tie the same weight to a 90 cm length of thread. Time 20 swings.
Tie the same weight to a 60 cm length of thread. Time 20 swings.
Does the length of the thread make any difference to the time of the swings?

3.

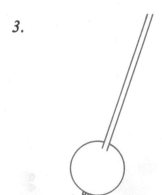

Here is the pendulum of a grandfather clock. Beneath the bob is a screw for lengthening or shortening the pendulum.
Think of the experiments you have been doing.
How would you adjust the clock if it was losing time ? Give the reason.
How would you adjust the clock if it was gaining time ? Give the reason.

Normally we use the 12-hour clock. The armed forces, shipping lines and airlines, however, always use the *24-hour clock*. This takes away any difficulties over a.m. and p.m. British Rail now uses the 24-hour clock, and soon this will be the way all time-tables will be written.

A This diagram shows the duty times of three watchmen.

Four numerals are always needed to write the time using the 24-hour clock system, like this :

6.5 p.m. becomes 18.05   3 p.m. becomes 15.00   1 a.m. becomes 01.00

Write the answers to these questions, using the 24-hour clock :
1. What time did Mr Tibbs finish his second duty ?
2. What time did Mr Jaggs start his first duty ?
3. What time did Mr Blake finish his second duty ?
4. What was the time half-way through Mr Jagg's second duty ?

B Here are some special 24-hour clocks showing the time in different parts of the world when it is 13.45 (or 1.45 p.m.) in London.

LONDON   NEW YORK   CALCUTTA   SINGAPORE

Write the above times, using the *12-hour* clock. Remember the minute hand goes round once every hour on both the 12-hour clock and the 24-hour clock.

C Here is part of a Green Line coach time-table. Write it out as it would appear in the 24-hour clock system.

		a.m.	a.m.	a.m.	a.m.	a.m.	a.m.		p.m.	p.m.	p.m.	p.m.
SLOUGH	*dep*.	9.11	9.41	10.11	10.41	11.11	11.41		12.11	12.41	1.11	1.41
HAMMERSMITH	*arr*.	9.54	10.24	10.54	11.24	11.54	12.24 (p.m.)		12.54	1.24	1.54	2.24

D Here is part of a South Wales time-table :

PADDINGTON *dep*. 17.40   CARDIFF *arr*. 19.55
NEWPORT *arr*. 19.41   SWANSEA *arr*. 21.05

Write out this time-table as it would have appeared before the Western Region changed to the 24-hour clock system.

A   Here are the departure times and flying times of some planes from London Airport. Work out the time of arrival of each plane, using Greenwich time.

	Dep. London	Destination	Flying Time	Time of Arrival
1.	09.55	Paris	55 min	☐
2.	13.15	Gibraltar	2 h 45 min	☐
3.	19.40	Athens	4 h 30 min	☐
4.	02.50	Hongkong	18 h 10 min	☐

B   You are given the Greenwich times of departure and arrival of some aircraft leaving London Airport. Work out the flying time for each one in hours and minutes.

	Dep. London	Destination	Time of Arrival	Flying Time
1.	23.30	Brussels	00.25	☐
2.	01.45	Tunis	04.25	☐
3.	17.05	Moscow	22.45	☐
4.	19.45	New York	02.25	☐

C

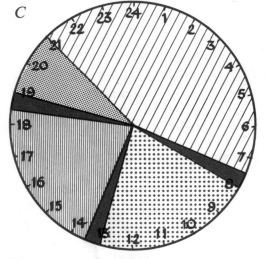

This pie chart shows how Tim spent the first 24 hours of his summer holiday. Answer the questions below, using the 12-hour clock.

1. At what time did the concert start ?

2. How long did the boat trip last ?

3. At what time did Tim start his lunch ?

4. How long did he spend in bed ?

5. How long did Tim spend playing on the beach ?

▧ IN BED   ■ MEALS   ▨ PLAYING ON BEACH   ▥ BOAT TRIP   ▦ CONCERT

Draw a similar pie chart and use it to show how you spent a day during your last holiday.

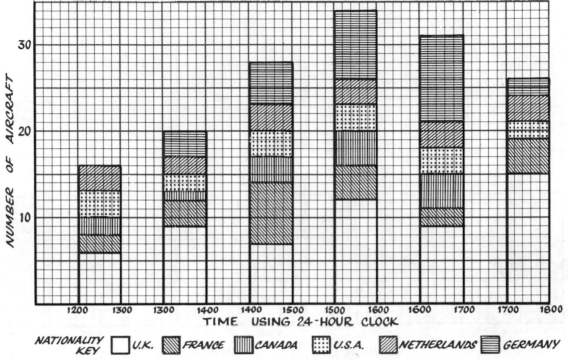

AIRCRAFT SPOTTING AT LONDON AIRPORT

NATIONALITY KEY ☐ U.K. ▨ FRANCE ▥ CANADA ⊞ U.S.A. ▧ NETHERLANDS ▤ GERMANY

This graph shows more information than the usual bar chart. It shows the total number of aircraft in each hour, and also the different nationalities of the aircraft.

A 1. How many aircraft were spotted between 14.00 hours and 15.00 hours ?
2. What was the greatest number of German aircraft spotted in an hour ?
3. How many more French than Canadian aircraft were seen between 14.00 hours and 15.00 hours ?
4. What was the total number of United States aircraft spotted altogether ?
5. Between 15.30 hours and 16.00 hours five United Kingdom aircraft were spotted. How many United Kingdom aircraft were spotted between 15.00 hours and 15.30 hours ?

B A group of children decided to make a count of the cars which passed their school. They chose four makes and then took turns to keep a careful tally.

Time	Austin	Vauxhall	Ford	Hillman
9–10 a.m.				
10–11 a.m.				
11–12 noon				
12–1 p.m.				

Make a traffic count of your own. Keep a careful tally and then draw up a bar chart to show the different makes of cars which passed in each hour.

LONDON AND WHIPSNADE ZOOS

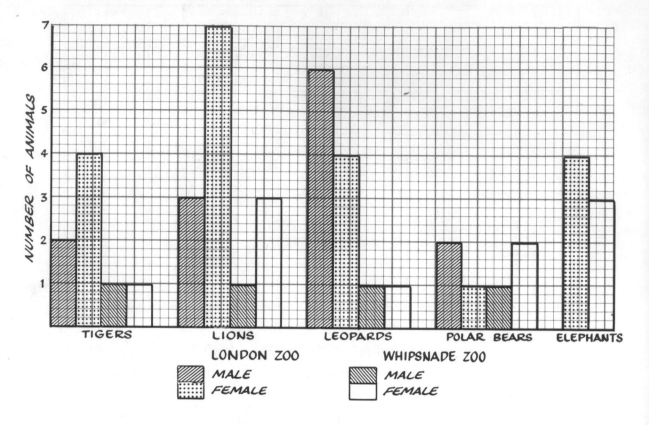

LONDON ZOO
MALE
FEMALE

WHIPSNADE ZOO
MALE
FEMALE

A   1. How many tigers and tigresses are there altogether in the two zoos ?
      2. Find how many more lionesses than lions there are in the two zoos.
      3. How many more male and female leopards are there than polar bears ?
      4. What is the total number of male and female tigers, lions and leopards at Whipsnade ?

B   This type of bar chart is very good for comparing examination results. Here are the marks obtained by four children preparing for their Cycling Proficiency tests.

	ANN	JILL	BOB	ALAN
First week	9	8	5	11
Second week	16	13	10	18

Draw a graph to show these results.

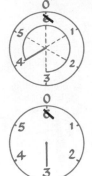

This special clock has only 1 hand and 6 numerals.
It is a '6-hour' clock.
If the hand is set at 4 and then moves 5 spaces, it will point to 3.

Now look at this example.

Set the hand on the numeral 3.

Move the hand 5 spaces.

$3+5 = 2$

A   Draw a 6 clock. Copy this table and then complete it.

Set hand on this number	4	5	5	4	3	2	2
Add this number	3	3	5	4	3	5	4
Sum of the 2 numbers	7	□	□	□	□	□	□
Time on the clock	1	□	□	□	□	□	□

What are the only numbers we need if we are using a 6 clock ?

B   Now try some additions on a 9 clock.

Draw a 9 clock and use it to complete this table.

First number	6	5	4	3	7	8
Add this number	8	7	5	7	4	8
Sum of 2 numbers	□	□	□	□	□	□
Time on the clock	□	□	□	□	□	□

What are the only numbers we use when working on a 9 clock ?
Can you see an easy way of calculating the numbers in the bottom line of the table ?

We can feel when the air is warm or cool and we can feel by touch when a thing is hot or cold.

However, this method is not accurate. To tell exactly how hot or cold a thing is we measure its TEMPERATURE. To do this we use a 'thermometer'.

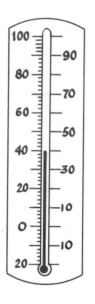

To find the temperature we read off the number of degrees.
This thermometer shows 40 degrees, or 40°.
The sign ° stands for degrees.
The markings at the side of a thermometer tube are known as 'the scale'.
This scale is called the CELSIUS scale.

60° stands for 60 degrees.
60° C stands for 60 degrees celsius.
What temperatures are shown by these thermometers?

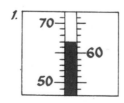

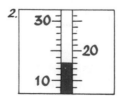

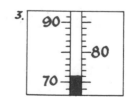

> You will need a celsius thermometer.

A Read the temperature of the room from your thermometer.
Each day you could keep a check on this and plot a graph to show what you have discovered.
Place your hand over the bulb at the bottom of the tube. What happens to the reading? Can you explain this?
Now place the thermometer under a cold water tap.
Note the temperature reading. Can you explain what has happened?

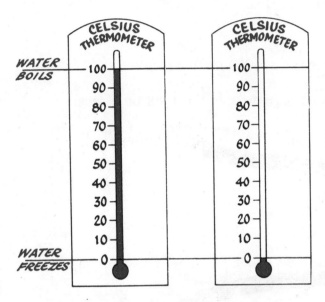

These two drawings show the temperature of water when it boils (100°) and when it begins to freeze (0°). These temperatures are known as boiling point (B.P.) and freezing point (F.P.).

B There are 100 degrees between the freezing point of water and its boiling point. Because of this the celsius scale is sometimes called a centigrade scale ('centigrade' means 'one hundred divisions'), but the temperature *can* go above and below these points. When it falls below 0° we say that it is below zero, and write it with a minus sign, *e.g.* 4° below zero is written −4°.

C What is the temperature on these thermometers?

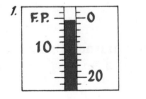
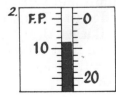

For very accurate measurement of short intervals of time, we use a stop watch. The dial is marked in seconds and some stop watches can record to $\frac{1}{10}$ of a second.

A   Look at these two stop watches. Write the time shown by each watch in minutes and seconds.

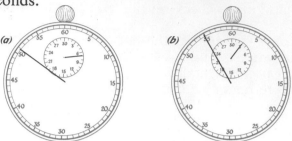

Which do you think shows the time for a 3 000 metres race?

B   Here are the stage times of the Cambridge crew in the 1948 boat race.

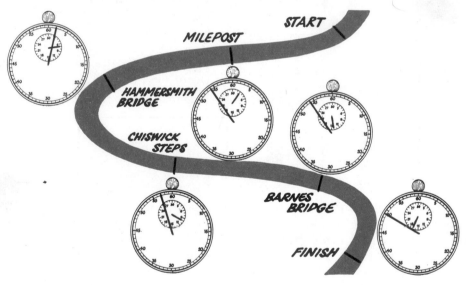

    1. What was the time for the whole course?
    2. What was the time in seconds for the first mile?
    3. How many minutes and seconds did the boat take from
        (*a*) Mile Post to Hammersmith Bridge?
        (*b*) Hammersmith Bridge to Chiswick Steps?
        (*c*) Chiswick Steps to Barnes Bridge?
    4. How many seconds did the boat take from Barnes Bridge to Finish?
    5. The record to Hammersmith Bridge is 6 min 45 sec, and was rowed by Cambridge in 1947. How many seconds more did the 1948 Cambridge crew take?

David on his tricycle could probably go about 6 kilometres in an hour.

John on his bicycle could travel about 20 kilometres in an hour.

Mike on his motor cycle could travel about 70 kilometres in an hour.

When we measure speed we must know two things :

    (*a*) the distance travelled ; (*b*) the time taken.

We usually talk about speed in 'kilometres per hour', that is, the number of kilometres travelled each hour ; but we could use other measurements for very slow speeds.

    A tortoise may go a distance of 10 centimetres in 1 minute : 10 cm per min.
    A hedgehog may go 10 metres in 1 minute : 10 m per min.

Can you think of other examples of speeds which would be too slow to measure in kilometres per hour?

John ran 1 kilometre in 5 minutes. We can turn this speed into kilometres per hour quite easily because we know that 5 minutes is $\frac{1}{12}$ of an hour.

        1 kilometre in 5 mins is the same as 12 kilometres per hour.

Do you think that John could run for a full hour at that speed?

A   Work out these speeds in kilometres per hour :

    1. 3 kilometres in 10 min   *2.* 50 kilometres in 6 min
    3. 7 kilometres in 12 min   *4.* 1$\frac{1}{2}$ kilometres in 20 min

    Which of these speeds do you think belongs to:

    (*a*) a cyclist? (*b*) a walker? (*c*) a car? (*d*) an aircraft?

B   Look at these pictures. Arrange them in order of speed, writing the letter of the fastest first.

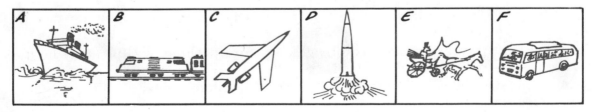

How fast do you walk?

> You will need a metre trundle wheel and a partner
> with a second hand to his watch.

When your partner with the stop watch signals that he is ready, walk around the perimeter of the playground with your trundle wheel. Count the number of revolutions until your partner signals that a minute is up.

You now have your speed in metres per minute.

How many metres can you walk in one hour?

Now change the metres to kilometres and you will know your speed in kilometres per hour (km/h).

Now *run* with the wheel and calculate your speed.

1. David rode his bicycle at a speed of 13 kilometres per hour. How far did he cycle in $2\frac{1}{2}$ hours?

2. An aircraft travelled at a speed of 700 kilometres per hour. How many hours did it take to fly 4900 kilometres?

3. A racing car made a time of $7\frac{1}{2}$ hours for a 675-kilometre race. What was the average speed of the car?

It is very difficult to keep to the same speed all the time.   A motorist has to reduce speed at bends and drive slowly through heavy traffic.   He is able to pick up speed on a clear, straight road.

A train has travelled 60 kilometres in an hour.   It may have been travelling at 5 or 10 kilometres per hour for part of the journey, and at 90 kilometres per hour at other times.   The *average* speed of the train for the whole way was 60 km/h. Look at its speeds as it passed through the stations on its 60-kilometre journey.

Distance	Station	Speed	Time
kilometres		km/h	min
0	DOKE	—	
20	BEWARK	82	
39	NETFORD	15	
44	BANSKILL	10	
56	LONCASTER	50	
58	BARKSEY	65	
60	MALTHOLME	—	60

A Bob recorded the time and kilometre reading of his father's car when they left Gloucester.

Bob also recorded the time and kilometre reading when they arrived at Taunton.

How long did the journey take?
What was the distance?
Work out the average speed in km/h.

B Now work out the TIME, DISTANCE or SPEED for each of these journeys.

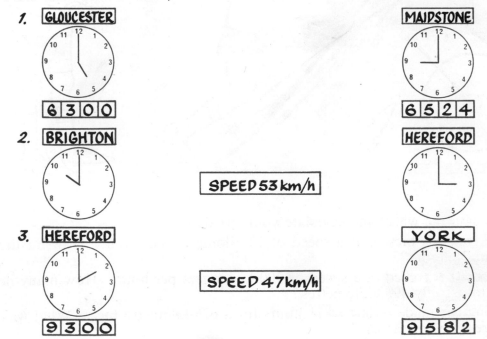

1. GLOUCESTER

6300

MAIDSTONE

6524

2. BRIGHTON

SPEED 53 km/h

HEREFORD

3. HEREFORD

SPEED 47 km/h

9300

YORK

9582

Look at these pictures carefully.   Make up a problem of your own for each picture and then show how you would solve the problem.

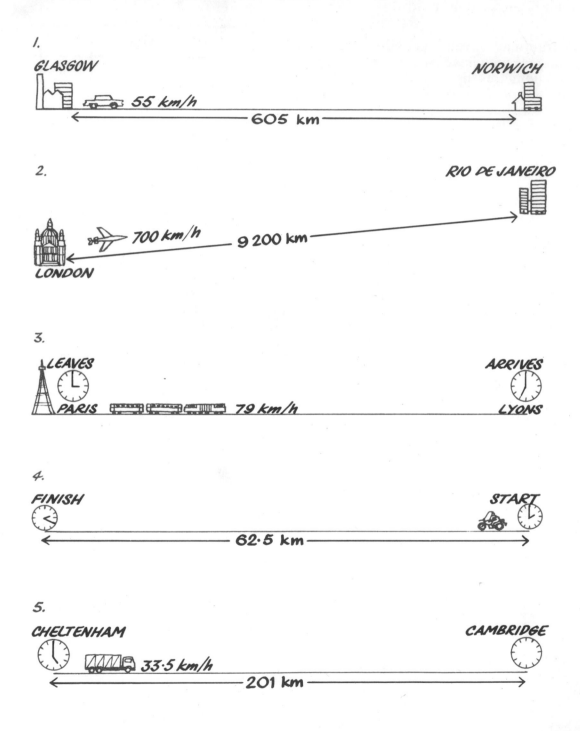

1.

GLASGOW   NORWICH

55 km/h

←———— 605 km ————→

2.   RIO DE JANEIRO

700 km/h   9 200 km

LONDON

3.

LEAVES   ARRIVES

PARIS   79 km/h   LYONS

4.

FINISH   START

←———— 62·5 km ————→

5.

CHELTENHAM   CAMBRIDGE

33·5 km/h

←———— 201 km ————→

We have learned that when we are counting today we group objects by tens. This is known as the *base ten* system.

If early man had learned to count using the fingers of one hand only, we might have had a number system based on sets of five. That would be a *base five* system.

(||||||||||) ||||||

1 ten and 6 units

We write: 16

(|||||) (|||||) (|||||) |

3 fives and 1 unit

We write: 31_{five}
We say: three one, base five.

A strange race of people with only four fingers would probably have used a *base four* system.

1   2   3 10_{four}

10_{four} stands for 1 four and 0 units.
We say: one zero, base four.

Here is a collection of marbles. Let us see how we can write the number in different bases.

After all possible sets of five have been made there are 3 left over. We write 33_{five}.

In the *base seven* system, we can see that there are two bases or sets of seven and four units left over. We write 24_{seven}.

1. This picture shows the same number of marbles collected into sets of six. How would you write this number in the *base six* system?

2. Each picture shows the same number of tally sticks collected into different sets. Write out the missing numerals.

 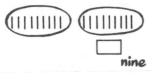 

five   nine   six

3. Here are some more tally marks:

/ /

Draw them on a piece of paper. Then group the marks into sets and write the number in *base nine*, *base eight*, and *base seven*.

A   Write the base of each numeral to show the number in each of these sets:

B   1. In base six, how many players are there in a cricket team?
      2. In base three, how many wheels are there on a pair of roller skates?
      3. In base five, how many eggs are there in two dozen?
      4. In base three, how many sides has a triangle?
      5. In base nine, how many are there in a score?

C   We now know that there are many ways of representing the same number by using different bases.

      Copy and complete this table.   The first line has been filled in for you.

	Base ten	Base eight	Base seven	Base nine	Base five	Base six
1.	17	21	23	18	32	25
2.	□	12	□	□	□	□
3.	□	□	20	□	□	□
4.	□	□	□	25	□	□

D   This plan shows the route from Bob's hut to his secret H.Q.   As you can see, the number of paces has been written in a special code for mathematicians.

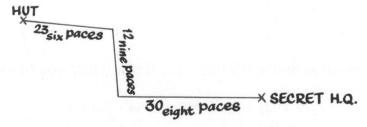

In base ten, work out the number of paces from the hut to the secret H.Q.

ORDER IN ADDITION

We know that the order in which we add two numbers makes no difference to the sum.

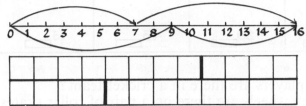

$$7+9 = 16$$
$$9+7 = 16$$

$$8+4 = 12$$
$$4+8 = 12$$

A   Remembering this law, solve these equations:

1. $9+\square = 6+9$ 2. $11+9 = 9+\square$ 3. $6+3\frac{1}{4} = 3\frac{1}{4}+n$
4. $398+37 = \square+398$ 5. $x+93 = \square+x$ 6. $\square+578 = 578+6{,}940$

GROUPING IN ADDITION

When we add, we can alter the grouping and get the same sum.

$$9+(2+8) = 19 \qquad (9+2)+8 = 19$$

B   Find these sums:

1. $13+4+6$ 2. $8+12+11$ 3. $23+7+17$ 4. $17+2+18$
5. $769+7+3$ 6. $680+20+239$ 7. $498+2+163$ 8. $95+5+673$

ORDER AND GROUPING IN ADDITION

We know from the two laws which tell us about order and grouping, that we can add any two numbers first and get the same sum.

$4+5+2$
$4+2+5$ 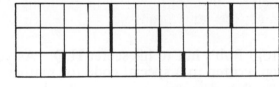   =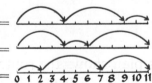
$2+5+4$

C   Remembering these two laws, find the sums:

1. $40+7+30+9$ 2. $70+43+30$ 3. $149+11+60+40$

D   Study the equation on the left first.   See if it will help you to solve the equation on the right.

1. $(79+94)+76 = 249$ $94+(79+76) = x$
2. $76+39 = 115$ $76+\square+39 = 155$
3. $36+14 = 50$ $36+97+14 = \square+50$

ORDER IN MULTIPLICATION

We know that the order in which we multiply two factors makes no difference to the product.

$$7 \times 3 = 21$$
$$3 \times 7 = 21$$

$$3 \times 4 = 12 \qquad 4 \times 3 = 12$$

A   Remembering this law, solve these equations:

1. $17 \times 4 = \square \times 17$ *2.* $19 \times \square = 11 \times 19$ *3.* $40 \times \mathbf{n} = 56 \times 40$

4. $76 \times 12 = 12 \times \square$ *5.* $\square \times 17 = 17 \times 96$ *6.* $60 \times 80 = 80 \times \mathbf{n}$

GROUPING IN MULTIPLICATION

When we multiply we can alter the grouping of the factors and get the same product.

$$(2 \times 3) \times 4 = 24 \qquad\qquad 2 \times (3 \times 4) = 24$$

B   Find these products:

1. $2 \times 5 \times 7$ *2.* $6 \times 10 \times 2$ *3.* $7 \times 10 \times 10$ *4.* $5 \times 2 \times 6$

C   Solve these equations:

1. $(4 \times 9) \times 2 = \square \times (9 \times 2)$ *2.* $19 \times (17 \times 16) = (\square \times 17) \times 16$

ORDER AND GROUPING IN MULTIPLICATION

We know, from the two laws which tell us about order and grouping, that we can multiply any two factors first and get the same product.

D   Find the products:

1.
$$6 \times 4 \times 3$$
$$6 \times 4 \times 3$$
$$6 \times 4 \times 3$$

2.
$$9 \times 2 \times 4$$
$$9 \times 2 \times 4$$
$$9 \times 2 \times 4$$

3.
$$8 \times 3 \times 5$$
$$8 \times 3 \times 5$$
$$8 \times 3 \times 5$$

4.
$$6 \times 2 \times 10 \times 5 \qquad 6 \times 2 \times 10 \times 5 \qquad 6 \times 2 \times 10 \times 5$$
$$6 \times 2 \times 10 \times 5 \qquad 6 \times 2 \times 10 \times 5 \qquad 6 \times 2 \times 10 \times 5$$

E   Find the products:

1. $4 \times 5 \times 7$ *2.* $5 \times 17 \times 2$ *3.* $8 \times 7 \times 5$ *4.* $25 \times 7 \times 4$

5. $10 \times 8 \times 10$ *6.* $5 \times 7 \times 4$ *7.* $10 \times 7 \times 9 \times 10$ *8.* $8 \times 7 \times 5 \times 1$

THE DISTRIBUTIVE LAW

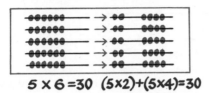

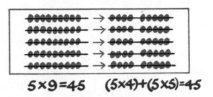

$$5 \times 6 = 30 \quad (5 \times 2) + (5 \times 4) = 30 \qquad 5 \times 9 = 45 \quad (5 \times 4) + (5 \times 5) = 45$$

> This law tells us that when we multiply we can 'split' a factor.

A Use these bead-frame pictures to complete the equations.

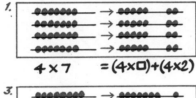

$$4 \times 7 \quad = (4 \times \square) + (4 \times 2) \qquad 3 \times 8 \quad = (3 \times 3) + (3 \times \square)$$

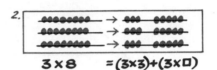

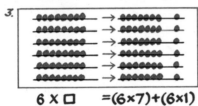

$$6 \times \square \quad = (6 \times 7) + (6 \times 1) \qquad 3 \times 7 \quad = (3 \times 4) + (3 \times \square)$$

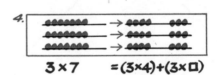

B Solve these equations:

1. $9 \times 8 = (5 \times 8) + (\square \times 8)$ 　　*3.* $\square \times 7 = (6 \times 7) + (5 \times 7)$
2. $6 \times 7 = (\square \times 2) + (6 \times 5)$ 　　*4.* $6 \times 9 = (2 \times 9) + (\square \times 9)$

C The pictures below show that it is possible to split either factor when we multiply.

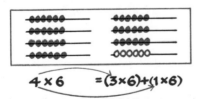

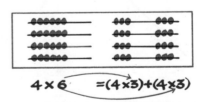

$$4 \times 6 \quad = (3 \times 6) + (1 \times 6) \qquad 4 \times 6 \quad = (4 \times 3) + (4 \times 3)$$

Solve these equations:

1. $4 \times 13 = (4 \times 9) + (4 \times \square)$ 　　　　*2.* $6 \times 27 = (6 \times 20) + (6 \times \square)$
3. $37 \times 4 = (30 \times 4) + (7 \times 4) = \square$ 　　*4.* $\square \times 9 = (8 \times 9) + (4 \times 9)$
5. $7 \times 85 = (7 \times 80) + (7 \times 5) = \square$ 　　*6.* $6 \times 12 = 60 + \square$

You will need card and scissors.

Many hundreds of years ago the Chinese discovered an interesting way of making shapes. You can do this by copying the square on to your card and cutting it into seven pieces.

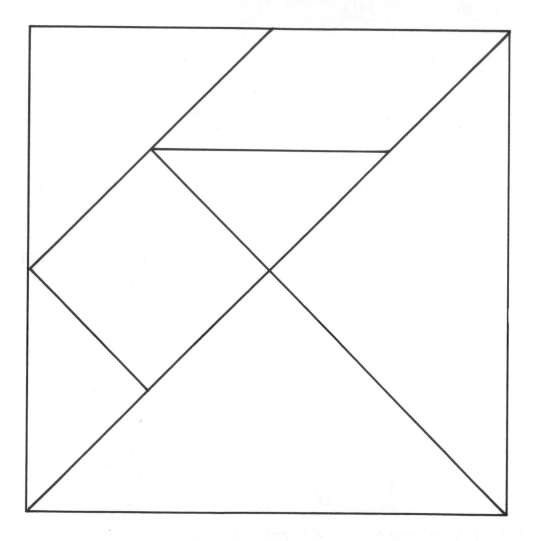

See how many different shapes you can make.
Copy them into your book.

ADDITION

A Find the sums:

1.	648	*2.*	43	*3.*	830	*4.*	987	*5.*	803
	+29		+768		+596		+696		+208

B Find the sums:

1.	9438	*2.*	6786	*3.*	72	*4.*	638	*5.*	688
	+ 89		+3913		140		45		33
					+ 89		+233		+ 7

6.	124	*7.*	473	*8.*	174	*9.*	6775	*10.*	7639
	365		19		706		9007		9634
	+945		+480		+399		+ 239		+2965

SUBTRACTION

C Find the answers and check by adding:

1.	764	*2.*	803	*3.*	970	*4.*	600	*5.*	805
	− 79		− 87		−188		− 33		−638

D Find the answers and check by adding:

1.	4703	*2.*	7605	*3.*	3038	*4.*	7000	*5.*	2016
	− 258		−4249		−1349		−5836		− 79

AVERAGES

E This table shows numbers for one day:

Class	Number in Class	Number Present	Milk	Dinners
1A	30	28	21	19
1B	35	34	30	30
2A	28	28	26	21
2B	31	30	24	15
3A	29	26	22	9
3B	33	28	27	26

1. Find the average number of pupils in a class.
2. Find the average number of pupils present in a class.
3. Find the average number of bottles of milk taken in a class.
4. Find the average number of dinners eaten in a class.

A

1858	7702	1488	2710	5373	510	1561	← LISBON
1448	9200	364	1155	5532	1244		← LONDON
1350	8099	1026	2358	5754			← MADRID
6850	7686	5825	5869				← NEW YORK
2005	10347	1315					← OSLO
1101	9120	← PARIS
9139	← RIO DE JANEIRO

ROME

Here is an air distance chart. To find the distance from Oslo to Lisbon, run your finger along the top row until it reaches the Oslo column. The square where the row and the column meet shows that the distance from Lisbon to Oslo is 2710 kilometres.

1. An aircraft flew from London to Paris and from Paris to New York. How many kilometres did it travel?
2. How much greater is the air distance from Lisbon to New York than from Lisbon to Paris?
3. (*a*) How much farther is the air distance from Madrid to Oslo than from Madrid to Paris?
    (*b*) How much farther is it from London to Rome than from Paris to Rome?
4. A plane flew from Rome to New York, from New York to London and from London to Oslo. How many kilometres were flown altogether?

B Write the correct symbol ($>$, $<$, $=$) in place of ● in each statement.

1. $93 + 86$ ● $76 + 96$ 2. $307 + 98$ ● $100 + 309$ 3. $900 + 70$ ● $700 + 90$
4. $540 + 60$ ● $450 + 100$ 5. $1000 - 1$ ● $900 + 100$ 6. $1000 - 25$ ● $500 + 250$
7. $100 - 1$ ● $1000 - 100$ 8. $1600 + 400$ ● $500 + 1500$ 9. $865 - 65$ ● $976 - 176$

C In each of these examples there are missing digits. Work out the missing digits and write out each example in full.

1.
```
    6 7 8
    4 7 7
  + □ 9 □
  ───────
  1 5 5 1
```
2.
```
    4 7 6
    □ □ □
  + 5 9 9
  ───────
  1 7 1 3
```
3.
```
  □ □ □ □
  + 6 7 5 9
  ─────────
  1 0 2 3 7
```
4.
```
    7 4 4 8
  − □ □ □ □
  ─────────
    5 8 7 9
```

5.
```
      9 □
  + □ 7
  ─────
  1 6 5
```
6.
```
      □ 8
  + 9 □
  ─────
  1 7 3
```
7.
```
    □ 9 □
  + □ 8
  ─────
  5 3 4
```
8.
```
  □ □ □ □
  − 3 4 9
  ───────
  6 7 3 5
```

$$3 \times 4 = 12 \begin{cases} 3 \text{ is a factor of } 12. \\ 4 \text{ is a factor of } 12. \end{cases}$$

$$12 \div 2 = 6 \begin{cases} 2 \text{ is a factor of } 12. \\ 6 \text{ is a factor of } 12. \end{cases}$$

3, 4, 2 and 6 are FACTORS of 12.
12 is a MULTIPLE of 3, of 4, of 2 and of 6.

We know that 7 divides into 21 leaving no remainder. 7 is thus a factor of 21.

A   Write true or false for each of these statements:

(*a*) 4 is a factor of 16.   (*b*) 3 is a factor of 20.   (*c*) 8 is a factor of 40.   (*d*) 5 is a factor of 29.   (*e*) 2 is a factor of 26.   (*f*) 1 is a factor of 9.

B   Here are four equations, each showing 48 written as the product of two factors:

$$1 \times 48 = 48$$
$$2 \times 24 = 48$$
$$3 \times 16 = 48$$
$$4 \times 12 = 48$$

(*a*) Write 48 as the product of two other factors.

(*b*) Take each of the following numbers and see in how many ways it can be shown as the product of two factors.

       16,   18,   30,   54

(*c*) List all the factors of these numbers: 28, 40, 36.

Every number has 1 as a factor and every number has itself as a factor.

C   If a number has only two factors, that is 1 and itself, it is a PRIME NUMBER. Copy these numbers and draw a ring around all the prime numbers:

20   27   29   31   36   39   43   49
51   57   69   72   81   91   96   100

D   The numbers 4, 6, 8, 10, 12, 14 and 16 are all multiples of the prime number 2. Which of the following numbers are multiples of the prime number 3?

7   6   9   15   20   29   39   40

A Greek mathematician called Eratosthenes, who lived about two hundred years before Christ, discovered a method of finding prime numbers by sifting away all the numbers that are not prime.

This method is called the 'Sieve of Eratosthenes'.

A   We can use the Sieve to discover all the prime numbers less than 30, like this:

1. Ring the prime number 2 and then cross out every second number. This means that we have crossed out all the multiples of 2.

2. Ring the next prime number, 3, and then cross out every third number. This means that we have crossed out all multiples of 3.

3. Ring the next prime number, 5, and then cross out every fifth number. This means that we have crossed out all multiples of 5.

4. Ring the prime number 7. Are there any multiples not crossed out? We have now sifted out all the numbers that are not prime. The numbers left in the Sieve are the prime numbers less than 30:

$$2, 3, 5, 7, 11, 13, 17, 19, 23, 29$$

B   Copy the list of numbers below. Use the Sieve to sift out all the numbers that are not prime numbers.

Now write down all the prime numbers remaining in the Sieve.

A   Study these multiplication examples carefully:

$$17 \times 11 = \frac{1\ 8\ 7}{1+7} \qquad 34 \times 11 = \frac{3\ 7\ 4}{3+4} \qquad 43 \times 11 = \frac{4\ 7\ 3}{4+3}$$

Can you write the products of the following examples without multiplying?

1. 27×11 *2.* 45×11 *3.* 16×11 *4.* 23×11

B   Copy these diagrams and write in the missing numbers:

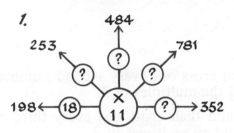   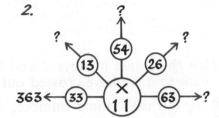

C   Copy this diagram and then write the following numbers in the circles around the triangle.   Use each number once only.

$\frac{1}{4}, \frac{1}{2}, \frac{3}{4}, 1, 1\frac{1}{4}, 1\frac{1}{2}, 2\frac{1}{2}, 3\frac{1}{4}, 4\frac{1}{4}$

The sum of the numbers on each side of the triangle must equal $6\frac{1}{2}$.
Five of the numbers have been filled in for you.

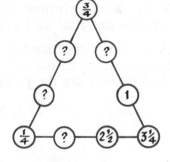

D   Write out these examples replacing □ or ● by a digit:

1.
```
  4 7 □
+1 2 8
------
  6 ● 1
```

2.
```
  9 3 6
- 1 □7
------
  ● 5 9
```

3.
```
   □7
  × 8
-----
 3 7 6
```

4.
```
      4 8
  9)4 □ □
    3 6
    ---
    7 2
    7 2
```

Now make up four examples of your own, putting question marks in place of the missing digits.

E   Think of a number, multiply it by 2, add 18, divide the result by 2 and subtract the number of which you first thought.   Try this several times, starting with a different number each time.   What do you notice about your answers?

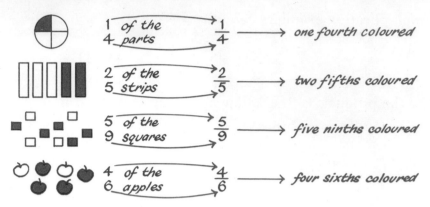

1. Write the missing numerals and words below.

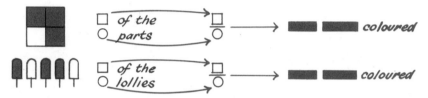

2. Study each picture below and work out the missing numerals.

	SHAPE	PARTS COLOURED	PARTS ALTOGETHER	FRACTION COLOURED
A	△	1	4	$\frac{1}{4}$ coloured
B		☐	7	$\frac{☐}{☐}$ coloured
C		4	☐	$\frac{☐}{☐}$ coloured
D		☐	☐	$\frac{☐}{☐}$ coloured

	SET	NUMBER COLOURED	NUMBER IN SET	FRACTION COLOURED
E		2	☐	$\frac{☐}{☐}$ coloured
F		☐	8	$\frac{☐}{☐}$ coloured
G		☐	☐	$\frac{☐}{☐}$ coloured
H		☐	☐	$\frac{☐}{☐}$ coloured

The numeral above the line in a fraction is called the NUMERATOR.
The numeral below the line is called the DENOMINATOR.

This picture shows a circle divided into 8 equal parts.
One part is coloured. $\frac{1}{8}$ of the circle is coloured.

1 is the NUMERATOR.
8 is the DENOMINATOR.

1. Which picture shows $\frac{1}{5}$ of a shape coloured?

2. Which picture shows $\frac{3}{8}$ of a shape coloured?

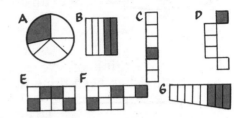

3. In this picture we can see that rod B is $\frac{4}{6}$ as long as rod A.

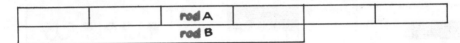

Look at these pictures and then write a fraction to compare the length of the
coloured rod with the length of the white rod.

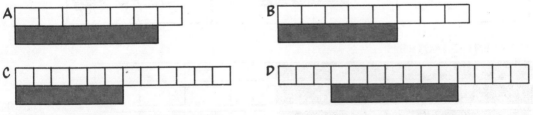

4. What fraction of AB is AD?

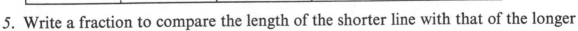

5. Write a fraction to compare the length of the shorter line with that of the longer
line.

A

B

C

D

6. What fraction of line AB is CD?

Rod A is divided into quarters.
Rod B is divided into halves.

We can see that $\frac{2}{4} = \frac{1}{2}$. Fractions that can be written equal to each other are called EQUIVALENT.

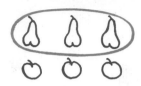

We can see that 3 of the 6 fruits, that is $\frac{3}{6}$, are pears.
We can also see that 1 of the 2 sets of fruits, that is $\frac{1}{2}$, are pears.
$\frac{3}{6}$ is equivalent to $\frac{1}{2}$

A   Can you think of two equivalent fractions to match each of these pictures? The first has been done for you.

1.
✱ ✱ ✱ ✱ ✱ ✱ ✱ ✱ $\frac{1}{2} \longrightarrow \frac{4}{8}$

2.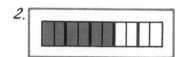

3.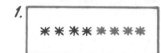

4.

5.

6. (dots image)

7.

8.

B   EQUIVALENT FRACTIONS

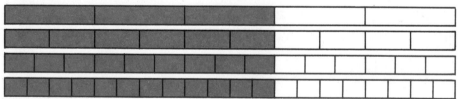

This fraction board shows the equivalent fractions $\frac{3}{5}, \frac{6}{10}, \frac{9}{15}, \frac{12}{20}$.
If we had a fraction board large enough, we could go on adding equivalent fractions:

$$\frac{3}{5}, \frac{6}{10}, \frac{9}{15}, \frac{12}{20}, \frac{15}{25}, \frac{18}{30}, \frac{21}{35} \cdots$$

Write the next two equivalent fractions in each of the following:

1. $\frac{2}{3}, \frac{4}{6}, \frac{6}{9}, \frac{8}{12}, \cdots$ *2.* $\frac{3}{8}, \frac{6}{16}, \frac{9}{24}, \frac{12}{32}, \cdots$

3. $\frac{5}{7}, \frac{10}{14}, \frac{15}{21}, \frac{20}{28}, \cdots$ *4.* $\frac{2}{9}, \frac{4}{18}, \frac{6}{27}, \frac{8}{36}, \cdots$

5. $\frac{4}{5}, \frac{8}{10}, \frac{12}{15}, \frac{16}{20}, \cdots$ *6.* $\frac{2}{10}, \frac{4}{20}, \frac{6}{30}, \frac{8}{40}, \cdots$

A *1.* Give three more fractions equivalent to $\frac{3}{7}$.

    2. Give three more fractions equivalent to $\frac{4}{9}$.

    3. Give three more fractions equivalent to $\frac{1}{10}$.

B Sometimes it is difficult to tell whether fractions are equivalent.   Here is a quick test:

$$\frac{3}{5} \diagdown\!\!\!\!\diagup \frac{6}{10} \quad \begin{array}{c} 30 \\ 30 \end{array}$$

This is called CROSS MULTIPLICATION.

    If the two products of the cross multiplication are the same number, the fractions are equivalent.

    Write the correct symbol, $=$ or $\neq$, between each pair of fractions.   If you are in doubt use the cross multiplication test.

1. $\frac{3}{7}$ and $\frac{15}{35}$ *2.* $\frac{3}{10}$ and $\frac{40}{400}$ *3.* $\frac{6}{8}$ and $\frac{30}{40}$ *4.* $\frac{5}{9}$ and $\frac{18}{21}$

5. $\frac{7}{8}$ and $\frac{14}{16}$ *6.* $\frac{42}{60}$ and $\frac{3}{5}$ *7.* $\frac{11}{12}$ and $\frac{33}{48}$ *8.* $\frac{10}{12}$ and $\frac{30}{36}$

C In the fraction $\frac{7}{8}$, we know that 7 is the *numerator* of the fraction and 8 is the *denominator*.   The numerator and denominator are called the TERMS of the fraction.

    $\frac{6}{12}$ is equivalent to $\frac{3}{6}$, but $\frac{3}{6}$ is in lower terms.

    $\frac{3}{6}$ is equivalent to $\frac{1}{2}$, but $\frac{1}{2}$ is in lower terms.

    $\frac{1}{2}$ cannot be reduced to lower terms.

    $\frac{1}{2}$ is therefore a fraction *in its lowest terms*.

1. Here is a set of equivalent fractions.   Write out the fractions with lower terms than $\frac{16}{20}$.

$$\left\{ \frac{4}{5}, \frac{64}{80}, \frac{8}{10}, \frac{32}{40}, \frac{80}{100} \right\}$$

2. Here is a set of equivalent fractions.   Write out the fractions with higher terms than $\frac{8}{24}$.

$$\left\{ \frac{1}{3}, \frac{32}{96}, \frac{40}{120}, \frac{56}{168}, \frac{2}{6}, \frac{12}{36} \right\}$$

3. Each of the following set of equivalent fractions has one fraction in its lowest terms.   Write out this fraction.

$$\left\{ \frac{70}{50}, \frac{21}{50}, \frac{42}{30}, \frac{63}{45}, \frac{7}{5}, \frac{35}{25} \right\} \qquad \left\{ \frac{24}{88}, \frac{21}{77}, \frac{3}{11}, \frac{6}{22}, \frac{15}{55}, \frac{18}{66} \right\}$$

A *1.* Write down the set of all fractions less than 1 which have (*a*) a denominator of 6; (*b*) a denominator of 10.

  2. Write down a set of eight equivalent fractions for each of these lowest term fractions: (*a*) $\frac{2}{5}$; (*b*) $\frac{5}{7}$; (*c*) $\frac{3}{8}$; (*d*) $\frac{4}{9}$.

  3. Write down the set of fractions which have a numerator of 3 and are greater than $\frac{3}{11}$. (The fractions must be less than $\frac{3}{3}$.)

B Write the correct sign, $<$, $>$ or $=$ between each pair of fractions:

  1. $\frac{3}{5}$ and $\frac{3}{10}$ *2.* $\frac{9}{100}$ and $\frac{7}{10}$ *3.* $\frac{2}{3}$ and $\frac{1}{3}$ *4.* $\frac{7}{8}$ and $\frac{7}{16}$

  5. Sort these fractions into two sets of equivalent fractions:

$$\frac{2}{3}, \frac{10}{15}, \frac{6}{10}, \frac{3}{5}, \frac{14}{21}, \frac{12}{20}, \frac{4}{6}, \frac{21}{35}, \frac{30}{50}, \frac{20}{30}$$

C When we reduce a fraction to its lowest terms we SIMPLIFY the fraction. Simplify these fractions:

  1. $\frac{30}{60}$ *2.* $\frac{28}{56}$ *3.* $\frac{48}{100}$ *4.* $\frac{7}{49}$ *5.* $\frac{99}{110}$ *6.* $\frac{20}{32}$ *7.* $\frac{48}{74}$

D What fraction is:

  1. 4 of a dozen?   *2.* 1 day of a week?

  3. 3 days of a week?   *4.* 5 of a score?

E Write these fractions in their lowest terms.

  What fraction is :

  1. 6 hours of a day?   *2.* 2 dozen of a gross?

  3. 30 of a century?   *4.* 10 minutes of an hour?

  5. 8 of 3 dozen?   *6.* 45 of 60?

F What fraction of each of these pictures is coloured?

1. 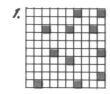   *2.* 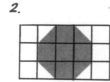   *3.*   *4.*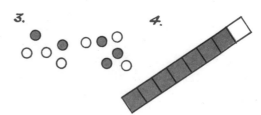

A   Study these drawings:

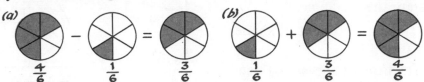

Now work these examples:

1. $\frac{5}{8} - \frac{3}{8} = \frac{\square}{8}$

2. $\frac{4}{11} + \frac{3}{11} = \frac{\square}{\bullet}$

3. We know $\frac{2}{9} + \frac{5}{9} = \frac{7}{9}$,
therefore   (*a*) $\frac{7}{9} - \frac{2}{9} = \frac{\square}{\bullet}$
and   (*b*) $\frac{7}{9} - \frac{5}{9} = \frac{\square}{\bullet}$.

4. We know $\frac{3}{8} + \frac{2}{8} = \frac{5}{8}$,
therefore   (*a*) $\frac{5}{8} - \frac{2}{8} = \frac{\square}{\bullet}$
and   (*b*) $\frac{5}{8} - \frac{3}{8} = \frac{\square}{\bullet}$.

B   How do we find the sum of $\frac{1}{3}$ and $\frac{1}{4}$?

We can change $\frac{1}{3}$ to twelfths: $\frac{4}{12}$

We can also change $\frac{1}{4}$ to twelfths: $\frac{3}{12}$

Now these two fractions can be added: $\frac{4}{12} + \frac{3}{12} = \frac{7}{12}$

We can, of course, find the sums of and differences between fractions without drawing pictures.
Look at this example: $\frac{1}{4} + \frac{2}{3}$

First we must think of other equivalent fractions for each. From these we choose fractions with the same denominator.

From our knowledge of sets of equivalent fractions, we know that:

$$\frac{1}{4} = \frac{3}{12} \text{ and } \frac{2}{3} = \frac{8}{12},$$
$$\text{therefore } \frac{1}{4} + \frac{2}{3} = \frac{3}{12} + \frac{8}{12} = \frac{11}{12}.$$

To find the sum of $\frac{1}{3} + \frac{1}{5}$, we change $\frac{1}{3}$ to $\frac{5}{15}$ and $\frac{1}{5}$ to $\frac{3}{15}$:

$$\frac{5}{15} + \frac{3}{15} = \frac{8}{15}$$

Try these: *1.* $\frac{1}{2} + \frac{1}{3} = \square$ *2.* $\frac{1}{4} + \frac{2}{5} = \square$ *3.* $\frac{5}{6} - \frac{1}{3} = \square$ *4.* $\frac{2}{3} - \frac{1}{4} = \square$

A 1. What fractional part of the set of shapes are the rectangles?

2. What fraction of the set of shapes are the triangles?

3. What fraction of the set of rectangles are the squares?

4. What fraction of the set of triangles are the right-angled triangles?

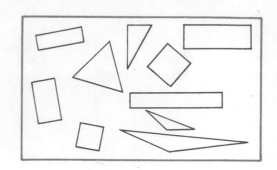

B

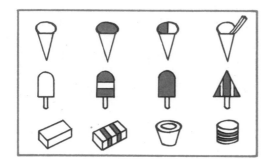

1. What fractional part of the set of ices are the cones? Can you write two fractions to show this?

2. What fractional part of the set of ice lollies is either striped or coloured?

3. What fractional part of the set of ices are the tubs? Can you write two fractions for this?

C In a magic square the numbers in each row, in each column and along each diagonal have the same sum.

Test the squares below to see if they are true magic squares:

$\frac{1}{3}$	$\frac{1}{8}$	$\frac{1}{6}$
$\frac{1}{24}$	$\frac{5}{24}$	$\frac{3}{8}$
$\frac{1}{4}$	$\frac{7}{24}$	$\frac{1}{12}$

$\frac{2}{3}$	$\frac{1}{12}$	$\frac{1}{2}$
$\frac{1}{4}$	$\frac{5}{12}$	$\frac{7}{12}$
$\frac{1}{3}$	$\frac{3}{4}$	$\frac{1}{6}$

Copy this magic square and fill in the missing numbers.

$\frac{1}{4}$	?	$\frac{1}{8}$
$\frac{3}{16}$	?	?
$\frac{1}{2}$	?	$\frac{3}{8}$

D Write out these statements, putting in the correct symbol (=, < or >) in place of ●:

1. $\frac{3}{4}$ ● $\frac{5}{8}$

2. $\frac{7}{8}$ ● $\frac{17}{24}$

3. $\frac{16}{24}$ ● $\frac{2}{3}$

4. $\frac{3}{5}$ ● $\frac{9}{10}$

5. $\frac{10}{20}$ ● $\frac{2}{3}$

6. $\frac{8}{12}$ ● $\frac{16}{24}$

7. $\frac{1}{2}$ ● $\frac{18}{36}$

8. $\frac{3}{4}$ ● $\frac{17}{24}$

A Copy and complete these tables.

1.

JILL HAD 50p	
Spent	*Amount Left*
20p	■
11p	■
■	35p
17p	■
■	33½p
10½p	■
39p	■
■	4½p

2.

BOB HAD £1	
Spent	*Amount Left*
30p	■
■	85p
63p	■
77p	■
■	8p
£0·36	■
■	£0·70
£0·03	■

3.

TIM HAD £5	
Spent	*Amount Left*
£3·10	■
£3·85	■
■	£4·37
£0·48	■
■	£1·72
£1·07	■
£2·16	■
■	£0·04½

4.

ANN HAD £10	
Spent	*Amount Left*
£7·50	■
■	£3·20
■	£7·10
£8·05	■
£0·15	■
£6·35	■
■	£7·19
£3·67	■

B Find the total amounts.

1. £
4·16
5·79

2. £
8·34
5·69

3. £
3·28
7·87

4. £
8·34
2·79

5. £
6·38
4·97
2·74

6. £
8·47
0·04
7·08

7. £
0·94
7·56
5·63

8. £
23·38
104·76
6·95

C Add these amounts. We must keep the decimal points in line so that we add pounds to pounds and new pence to new pence.

1. £8·94, £11·95, £0·83
2. £34·37, £7·10, £0·77
3. £57·96, £0·40, £10·09
4. £100·83, £0·75, £9·09

D Find the difference in the amounts.

1. £
5·67
3·49

2. £
34·84
14·67

3. £
8·76
3·07

4. £
7·04
2·46

5. £20·00, £7·34
6. £37·66, £60·04
7. £106·36, £39·87
8. £7·96, £23·48
9. £135·27, £18·69
10. £3·78, £62·40

A   We sometimes have a multiplication example
with carrying, like this :

$$\begin{array}{r} £0·71 \\ \times 6 \\ \hline £4·26 \end{array}$$

Now try these examples :

1. £0·42	*2.* £0·64	*3.* £0·49	*4.* £0·58
×4	×3	×4	×5

B   When we multiply an amount that is more than one pound,
we multiply like this :

$$\begin{array}{r} £4·03 \\ \times 2 \\ \hline £8·06 \end{array}$$

Now try these examples :
We must make sure that we put the decimal point in the right place.

1. £2·08½	*2.* £5·27	*3.* £5·27	*4.* £9·70
×6	×4	×7	×9
5. £4·57	*6.* £17·39	*7.* £20·30	*8.* £47·17
×8	×6	×7	×8
9. £7·09½	*10.* £24·90	*11.* £0·94½	*12.* £111·47
×9	×6	×8	×6

C   We divide money numbers in the same
way as we divide other numbers. We
must be careful to place the decimal point
in the quotient above the decimal point
in the dividend.

$$\begin{array}{r} £1·36 \\ 4)\overline{£5·44} \\ 4 \\ \hline 14 \\ 12 \\ \hline 24 \\ 24 \end{array} \quad \begin{array}{r} \text{check} \\ £1·36 \\ \times 4 \\ \hline 5·44 \end{array}$$

1. Five boys shared the cost of buying a
microscope costing £8·25. What was
each boy's share?

2. 6)£44·94	*3.* 8)£25·12	*4.* 9)£12·51	*5.* 8)£75·52
6. 7)£28·42	*7.* 8)£77·60	*8.* 4)£53·36	*9.* 8)£61·12
10. 5)£46·50	*11.* 7)£66·43	*12.* 6)£17·82	

1. Work out the cost of a tennis racket and 2 dozen tennis balls.

2. If you handed the shopkeeper three £5 notes to pay for 2 footballs, what change would you have?

3. I received £7·54 change from a £10 note. What had I bought?

4. Bob has already saved £5·76. How much more must he save to buy a football and a pair of football boots?

5. The skis cost how much more than the punchball and a cricket bat together?

6. Tim spent exactly £7. Which two articles did he buy?

7. What is the total cost of the three most expensive articles?

8. How much more must be paid for a dozen cricket balls than a dozen tennis balls?

We have already learned that to compare fractions we can think of equivalent fractions with the same denominator:

$$\frac{3}{4} = \frac{18}{24} \qquad \frac{2}{3} = \frac{16}{24}$$

Fractions can also be compared by bringing them to hundredths:

$$\frac{1}{20} = \frac{5}{100} \qquad \frac{1}{50} = \frac{2}{100} \qquad \frac{7}{10} = \frac{70}{100}$$

A

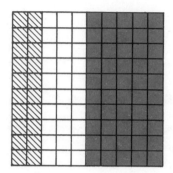

Look at this large square divided into 100 smaller squares.

1 small square is $\frac{1}{100}$ of the large square.

1. How many hundredths are there in $\frac{1}{2}$ of the large square?

2. How many hundredths are there in $\frac{3}{10}$ of the large square?

3. How many hundredths are there in $\frac{1}{5}$ of the large square?

We can see from our diagram that $\frac{1}{10} = \frac{10}{100}$.

We mean that 1 part out of 10 is the same as 10 parts out of 100.

Instead of writing 10 parts out of a 100 as a fraction, an easier way to show fractions with a denominator of 100 can be used.

We can write 10 parts out of 100 as a PERCENTAGE.

We say ' ten per cent '. We write 10%.

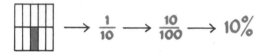

$$\longrightarrow \frac{1}{10} \longrightarrow \frac{10}{100} \longrightarrow 10\%$$

B   Find the missing symbols:

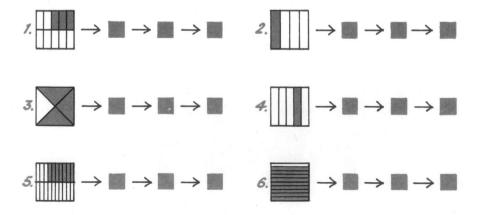

> By 'per cent' we mean 'per hundred' or 'parts of a hundred'.

A   Here is a tray holding 100 beads.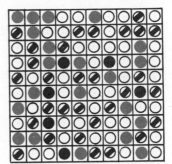

1. What percentage of the beads are 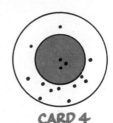 ?
2. What percentage of the beads are ● ?
3. What percentage of the beads are ○ ?
4. (*a*) What percentage of the beads are ◐ ?
    (*b*) Write this as a fraction in its lowest terms.
5. Add up all the percentages.

B   Bob made some very simple target cards for practice with his pea-shooter. On card 1 he scored 4 bull's-eyes out of 16 tries. What percentage of the peas scored bull's-eyes on card 1?

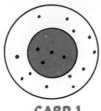 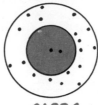 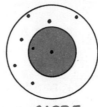

CARD 1   CARD 2   CARD 3   CARD 4

He first expressed the number as a fraction ⟶ $\frac{4}{16}$;
then as a fraction in its lowest terms ⟶ $\frac{1}{4}$;
then as a fraction with a denominator of 100 ⟶ $\frac{25}{100}$;
then as a *percentage* ⟶ 25%.

Work out the percentage of bull's-eyes scored on card 2, card 3 and card 4.

C   Record these as percentages:

1. 8 out of 40   2. 30 out of 60   3. 17 out of 50
4. 40 out of 50   5. 6 out of 30   6. 70 out of 100

D   Change each fraction to a fraction with a denominator of 100, and then write as a percentage:

$\frac{3}{50}$ $\frac{9}{20}$ $\frac{3}{10}$ $\frac{2}{5}$ $\frac{4}{25}$ $\frac{17}{20}$

E   Write these percentages as fractions in their lowest terms.

60%   90%   75%   35%   50%   8%

You will need a coin.

Toss a coin and see whether it comes down 'heads' or 'tails'.

You can keep count in this way:

HEADS	|||| |||| ||
TAILS	|||| |||

Toss 100 times and count the number of heads and the number of tails. Are the two numbers about equal?
The coin can land in 1 of 2 ways.
We say there is 1 chance in 2 of getting 'heads'.

1. What is the chance of getting 'tails'?

2. If you tossed the coin 20 times, how many 'heads' and how many 'tails' would you expect to get?

Bob has a picture card. If he flicked it into the air it would land showing either the picture side or the blank side.
The card would land in one of two ways:

We say there is 1 chance in 2 of the picture side showing. We usually write this in fraction form, like this: $\frac{1}{2}$.

Which hand is it in? Three boys hold out their clenched fists; in one of the hands there is a small pebble. What is the chance of choosing the hand with the pebble?

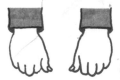 

There are six possible choices, but only one will be the right choice. The chances of guessing correctly on a single choice are 1 out of 6. This can be shown in fraction form: $\frac{1}{6}$.

If there are two pebbles, one in each of two hands, the chances of choosing a correct hand will be 2 out of 6, $\frac{2}{6}$, or in its simplest form, $\frac{1}{3}$.

A

1. Look at this top. What is the chance of getting a 6, a 2 or a 3?

2. What is the chance of getting either a 4 or a 5?

Below are three 'lucky number' cards and you are allowed to choose one number on each.

(a)

1	2	3	4	5	6
7	8	9	10	11	12
13	14	15	16	17	18

(b)

1	2	3	4	5	6
7	8	9	10	11	12
13	14	15	16	17	18
19	20	21	22	23	24

(c)

1	2	3	4	5	6	7	8
9	10	11	12	13	14	15	16
17	18	19	20	21	22	23	24
25	26	27	28	29	30	31	32
33	34	35	36	37	38	39	40

On this card there are 2 lucky numbers

On this card there are 3 lucky numbers

On this card there are 4 lucky numbers

3. What is the chance of winning on card (a), card (b), card (c)?

4. Which card offers the best chance of winning?

B   Twelve slips of paper are placed in a hat. Three of the slips are marked with a cross. What is the chance of drawing out a slip with a cross on the first try?

We know there are 12 possible choices.
We know there are 3 slips with crosses.
The chances of drawing out a slip with a cross is $\frac{3}{12}$.
Written in its simplest form, the fraction is $\frac{1}{4}$.

Here is the top layer of a box of chocolates. If I picked up one without looking, what would be the chance of:

1. Choosing a plain chocolate?
2. Choosing a caramel?
3. Choosing a Turkish delight?

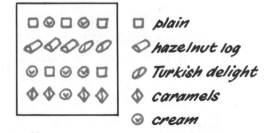

Now try to make a work card by thinking up some problems of your own.

Here is a plan showing the position of all the desks in a classroom.

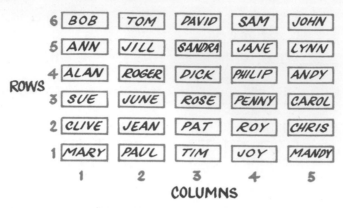

ROWS

	1	2	3	4	5
6	BOB	TOM	DAVID	SAM	JOHN
5	ANN	JILL	SANDRA	JANE	LYNN
4	ALAN	ROGER	DICK	PHILIP	ANDY
3	SUE	JUNE	ROSE	PENNY	CAROL
2	CLIVE	JEAN	PAT	ROY	CHRIS
1	MARY	PAUL	TIM	JOY	MANDY

COLUMNS

The set of pupils in column 3 is Tim, Pat, Rose, Dick, Sandra, David.
The set of pupils in row 2 is Clive, Jean, Pat, Roy, Chris.
Only Pat is in both column 3 *and* row 2.
We can describe Pat's position as (3, 2).
Is this the same as (2, 3)?
Tim's position is (3, 1).
Does it matter in which order we write the numerals?

Here are boxes of model cars stacked neatly on a shop shelf.

	1	2	3	4	5	6	7	8	9
5	JAGUAR	AUSTIN MINI	BUICK	CHRYSLER	ROVER	CORTINA	LINCOLN	ZEPHYR	REFUSE TRUCK
4	OPEL	PONTIAC	VICTOR	ROLLS ROYCE	CORSAIR	FIAT	M.G.1100	ANGLIA	TRACTOR
3	IMP	LOTUS	RENAULT	DAIMLER	BENTLEY	TROJAN	AUSTIN 1800	CITROEN	FORD G.T.
2	ASTON MARTIN	M.G.B. SPORTS	VOLVO	TRIUMPH 2000	MORRIS 1000	HILLMAN IMP	SAAB 96	ZODIAC	A.A. VAN
1	B.R.M.	COOPER	FERRARI	DUMPER	B.P. TANKER	CATTLE TRUCK	POLICE CAR	WRECK TRUCK	FIRE ENGINE

1. We can record the position of the Fiat as (6, 4).
Record the position of (*a*) the Daimler
(*b*) the Jaguar
(*c*) the Zodiac.

2. Which models are in these positions:
(*a*) (7, 3)?   (*b*) (5, 4)?   (*c*) (8, 4)?
(*d*) (3, 1)?   (*e*) (4, 4)?   (*f*) (7, 5)?

A

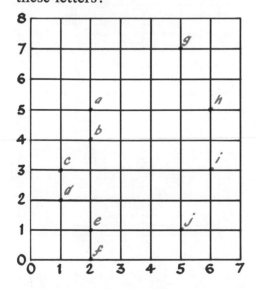

Look at this block of flats:

1. (*a*) In which window is there a cat?

   (*b*) Which window has a flower box?

2. (*a*) What is the position of the window with a notice?

   (*b*) What is the position of the window with a vase of flowers?

B This diagram shows counters which have been arranged on a sheet of graph paper.
We can describe the position of each counter very accurately, for the lines have been numbered.
We can describe the position of the dotted counter as (3, 1). Write a number pair to describe the position of (*a*) the white counter; (*b*) the coloured counter; (*c*) the lined counter.

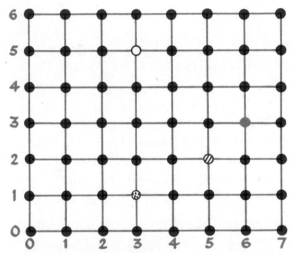

C Write the number pair for each of these letters:

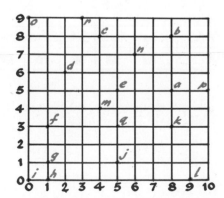

D Give the letter for each of these number pairs.

1. (2, 6)   *2.* (5, 1)   *3.* (8, 8)
    (5, 5)   (1, 1)   (2, 6)
    (0, 9)   (9, 0)   (4, 4)
    (10, 5)   (6, 7)   (3, 9)

A

A long-distance cyclist travelled at an average speed of 20 kilometres per hour. Bob made this graph to show the relationship between the number of hours and the distance travelled.

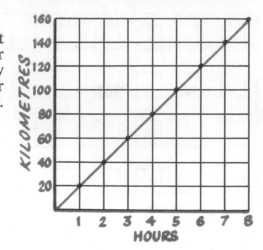

Bob first drew his horizontal axis and marked it off in hours. You can see that each division on the horizontal axis represents 1 hour. This is known as the *horizontal scale*.

Next, Bob drew a vertical axis and marked the kilometres on it. You can see that each division on the vertical axis represents 20 kilometres. This is known as the *vertical scale*.

Bob then plotted points to show that in 1 hour the cyclist travelled 20 km, in 2 hours he travelled 40 km and so on.

What do you notice about the position of the points plotted?

Do we need to plot all the points if the graph is a straight line?

Use the graph to answer these questions :
  1. How many kilometres were travelled in $3\frac{1}{2}$ hours?
  2. How long did the cyclist take to travel 90 km?

B Bob drew this graph to show the number of pencils he sharpened in a certain time. Copy the table below. Then use the graph to complete it.

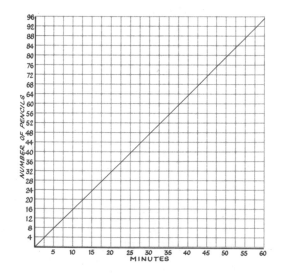

TIME	5 min	10 min	$\frac{1}{4}$ hr	20 min	25 min	$\frac{1}{2}$ hr
PENCILS	■	■	■	■	■	■

How long did Bob take to sharpen 12 pencils?

In the early days of sea travel the speed of a ship was calculated by measuring the distance travelled in a certain time.

For this purpose a special line called the 'log line' was thrown overboard. The line was knotted at equal distances and the end was attached to a log to keep it afloat. A sailor counted the number of knots that ran out as the ship sailed on its course.. So it was possible to work out the distance travelled in a certain time.

Today the speed of a ship is recorded by instruments, but the unit of speed is still called a KNOT.

A knot means one nautical mile per hour. So we can say that the speed of a ship is '23 knots'. We never speak of 'knots per hour'.

CONVERSION GRAPH
KNOTS TO KILOMETRES PER HOUR : KILOMETRES PER HOUR TO KNOTS

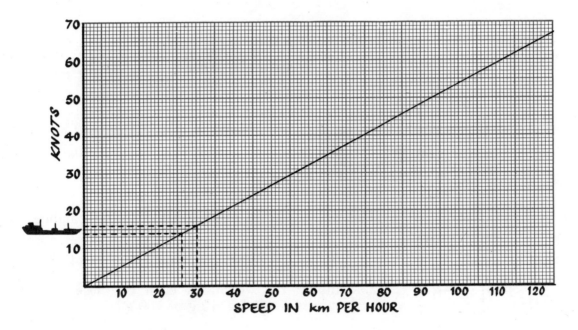

1. What approximately is a speed of 30 km per hour in knots?
2. Find the speed of the tanker in km per hour.

Now write a work card for your friends by making up problems of your own, using the graph and the information below.

HYDROFOIL PATROL BOAT, 66 knots	FAST DESTROYER, 45 knots
'QUEEN ELIZABETH II' (normal speed), 28 knots	PADDLE STEAMER, 17 knots
AIRCRAFT CARRIER, 36 knots	HOVERCRAFT, 60 knots

GRAMMES

The unit of weight in the Metric System is the GRAMME (g).

A   Jill weighed some of the objects in her satchel.

135 g   35 g   5 g   15 g

1. What is the total weight of these things in grammes?
2. How much heavier is the pear than all the other things together?
3. How many of the rubbers would be needed to be the same weight as the pear?

TENS OF GRAMMES

B

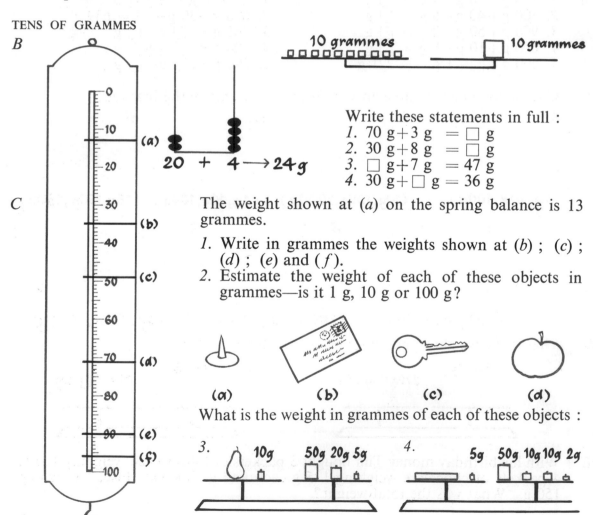

Write these statements in full :
1. 70 g + 3 g = ☐ g
2. 30 g + 8 g = ☐ g
3. ☐ g + 7 g = 47 g
4. 30 g + ☐ g = 36 g

C   The weight shown at (*a*) on the spring balance is 13 grammes.

1. Write in grammes the weights shown at (*b*) ; (*c*) ; (*d*) ; (*e*) and (*f*).

2. Estimate the weight of each of these objects in grammes—is it 1 g, 10 g or 100 g?

(a)   (b)   (c)   (d)

What is the weight in grammes of each of these objects :

3.   10g   50g 20g 5g   *4.*   5g   50g 10g 10g 2g

HUNDREDS OF GRAMMES

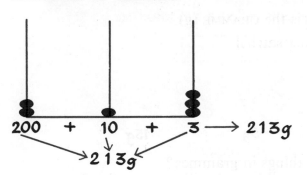

$200 + 10 + 3 \longrightarrow 213g$

$213g$

The weight of 100 GRAMMES

 in iron

 in brass

Write out the number which can replace □ in each of these statements :

A 1. 100 g + 20 g + 7 g = □ g   *B* 1. □ + 70 g + 8 g = 578 g
    2. 300 g + 40 g + 6 g = □ g   2. 600 g + 30 g + □ = 639 g
    3. 900 g + 50 g + 2 g = □ g   3. 361 g = □ g + □ g + □ g
    4. 600 g + 10 g + 8 g = □ g   4. 708 g = □ g + □ g
    5. 200 g + □ + 1 g = 221 g   5. 390 g = □ g + □ g

C Choose the most suitable answer from the weights in the brackets.

1.   *2.*   *3.*   *4.*

(1g, 10g, 100g)   (10g, 100g, 1000g)   (1g, 10g, 100g)   (15g, 150g, 1500g)

5.   *6.*   *7.*

(600g, 60g, 6g)   (200g, 20g, 2g)   (340g, 34g, 3·4g)

D What is the weight in grammes of each of these parcels?

1. 10g 2g   200g 100g 50g   *2.* 20g 5g   100g 50g 10g

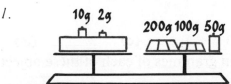   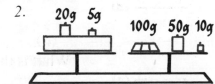

E With his birthday money Tim bought 3 packets of toffees each weighing 100 g, 4 packets of chocolate weighing 125 g each and a chocolate biscuit weighing 150 g. What was the total weight?

THOUSANDS OF GRAMMES

The weight of 1000 GRAMMES is called a KILOGRAMME

$$1 \text{ KILOGRAMME} \leftrightarrow 1000 \text{ GRAMMES}$$

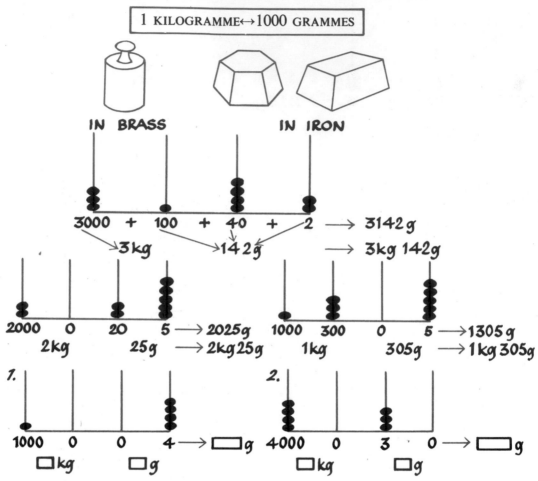

IN BRASS   IN IRON

3000 + 100 + 40 + 2 ⟶ 3142 g
⟶ 3 kg ⟶ 142 g ⟶ 3 kg 142 g

2000 0 20 5 ⟶ 2025 g   1000 300 0 5 ⟶ 1305 g
2 kg   25 g ⟶ 2 kg 25 g   1 kg   305 g ⟶ 1 kg 305 g

1. 1000 0 0 4 ⟶ ☐ g   2. 4000 0 3 0 ⟶ ☐ g
☐ kg ☐ g   ☐ kg ☐ g

Write out these statements in full :

3. 4000 g + 700 g + 30 g + 5 g → ■ g → ■ kg ■ g
4. 3000 g + 40 g + 6 g → ■ g → ■ kg ■ g
5. 1000 g + 200 g + 2 g → ■ g → ■ kg ■ g
6. 6000 g + 70 g → ■ g → ■ kg ■ g
7. 9000 g + 900 g → ■ g → ■ kg ■ g
8. What are the weights in grammes of these parcels?

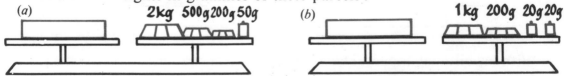

(a)   2kg 500g 200g 50g   (b)   1kg 200g 20g 20g

9. A full tin of toffees weighs 1 kg 100 g. The empty box weighs 225 g. What is the weight of the toffees in grammes?

The LITRE is the standard unit of capacity in the metric system.

My bucket holds
about 10 litres.

My watering can
holds about 6 litres.

Great use is also made of the DEMI-LITRE or half litre.

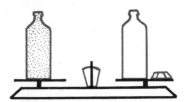

The QUARTER LITRE is also a useful measure.
We know that a litre of water weighs a kilogramme.

A *1.* A bottle weighs 650 g and holds 2·5 litres of water. What is the total
weight in grammes?

2. 1½ litres of water was just enough to fill ten small glasses. What, in grammes,
was the weight of water in each glass?

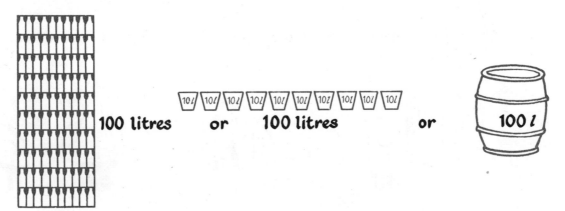

100 litres   or   100 litres   or   100 *l*

B *1.* How many double litres of wine are there in the full barrel?

2. An empty 10-litre bucket weighs 3·5 kg. What is the total weight of the
bucket, in grammes, when it is half full of water?

3.

(a) When the petrol tanker was half full
of petrol, how many litres did it hold?

(b) What is the weight of the petrol when it is full? A litre of petrol
weighs 750 g.

Greatest distances are measured in metres and kilometres.

A

```
        A 46
Lincoln      19 km
Newark       45 km
Nottingham   77 km
Leicester   101 km
```

1. A surveyor's chain measures 10 metres. How many lengths of the chain measure a quarter of a kilometre?

2. A square field has sides of 300 m. What is the perimeter of the field in kilometres and metres?

3. Write out these statements in full :
   (*a*) 1350 m + ☐ m = 1·5 km
   (*b*) 1760 m + 740 m = ☐ km

B   Many measurements of length are expressed in metres and in centimetres.

1. From a 20-metre length of wire I cut 11 m 50 cm, 7 m 20 cm and 1 m 10 cm. What length of wire was left?

2. A square room has sides of 4 m 65 cm. What is the perimeter of the room in metres and centimetres?

3. Bob's father is 1 m 85 cm tall. Bob is 119 cm tall. How much taller is Bob's father?

4. Write out these statements in full :
   (*a*) 199 cm + ☐ cm = 2·5 m
   (*b*) ☐ cm + 171 cm = 2 m

Small things can be measured in centimetres.

Very small things can be measured in MILLIMETRES (mm).

10	20	30	40	50	60	70	80	90	100	← *millimetres*
1	2	3	4	5	6	7	8	9	10	← *centimetres*

Each centimetre is divided into 10 millimetres.

A Here are the lengths of some model cars in millimetres. Write the lengths in centimetres and millimetres. The first has been done for you.

MG 1000	66 mm	6 cm 6 mm
Rolls Royce	72 mm	■ cm ■ mm
Ford Anglia	50 mm	■ cm ■ mm
Lotus Climax	91 mm	■ cm ■ mm

B Measure these lines carefully. Write the lengths in centimetres and millimetres, and then in millimetres.

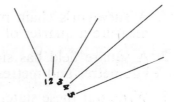

C

A **METRE** is the standard unit of length

A **CENTIMETRE** is one-hundredth of a metre ($\frac{1}{100}$)

A **MILLIMETRE** is one-thousandth of a metre ($\frac{1}{1000}$)

The length of this line is :

       123 millimetres (123 mm) or

       12 centimetres 3 millimetres (12 cm 3 mm)

This table gives the lengths of some model toys. Copy the table and complete it.

Tractor and trailer	242 mm	24 cm 2 mm
Twin tippers	302 mm	■ cm ■ mm
Breakdown truck	121 mm	■ cm ■ mm
Double freighter	■ mm	■ cm ■ mm
Car transporter	■ mm	20 cm 9 mm

D Write out these sentences, putting in the correct word—metre, centimetre or millimetre :

1. This page is about 25 ☐ long.

2. The packet of postcards is about 6 ☐ thick.

3. The garden is about 80 ☐ long.

4. The door is about 2 ☐ high and 1 ☐ wide.

A box of brass weights holds :

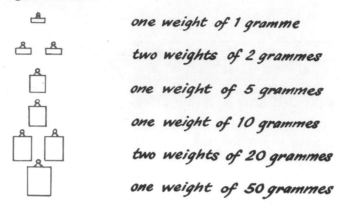

one weight of 1 gramme

two weights of 2 grammes

one weight of 5 grammes

one weight of 10 grammes

two weights of 20 grammes

one weight of 50 grammes

The shopkeeper does not often use small weights. He will seldom need weights of less than 50 grammes.

His weights are usually made of iron.

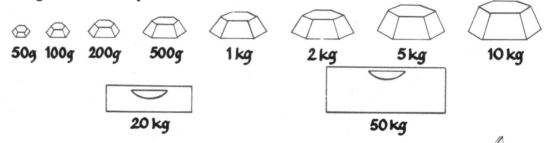

50g   100g   200g   500g   1kg   2kg   5kg   10kg

20 kg   **50kg**

A   1. When do you think the 50 kilogramme weight would be used? Perhaps this drawing will give a clue?

    2. Make up these weights with the least number of weights possible :
      (*a*) 225 g   (*b*) 480 g   (*c*) 810 g

THE METRIC TONNE

METRIC TONNE ←→ 1000 KILOGRAMMES

B   1. Write in tonnes (t) and kilogrammes: 2850 kg, 7060 kg, 3100 kg.
    2. Write in kilogrammes : 6 t 70 kg, 3 t 410 kg, 6 t 9 kg.
    3. A lorry can carry a load of 6 metric tonnes. How many bars of iron each weighing 40 kg can be carried?
    4. What is the weight in metric tonnes of 8 large crates each weighing 750 kg?

Weight of full tin 270 g→gross wt.
Weight of empty tin 43 g
Weight of beans 227 g→net wt.

1.

(a) What is the gross weight?
(b) What is the weight of the container?
(c) What is the net weight?

2.

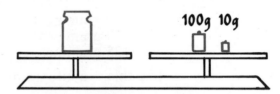

What is the weight of the mincemeat?

3.

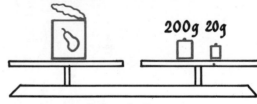

What is the weight of the contents of the tin?

4.

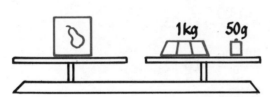

What is the weight of the sardines?

5.

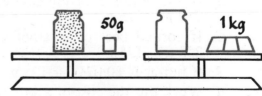

What is the weight of the jam?

A

10
metres of ribbon cost £0·40
metres of rayon cost £1·10
metres of braid cost £0·35
litres of wine cost £11·5
kilogrammes of apples cost £1·40

1
metre costs   £ ■
metre costs   £ ■
metre costs   £ ■
litre costs   £ ■
kilogramme costs £ ■

B

100
bulb holders cost £4·50
torch batteries cost £6·00
magnets cost £15·00
electric buzzers cost £20·00
electric motors cost £123·00

1
costs £ ■
costs £ ■
costs £ ■
costs £ ■
costs £ ■

When we know the price of a kilogramme it is easy to find the price of

> 500 grammes
> 250 grammes
> 125 grammes

like this : a kilogramme of ham cost £1·00

$$500 \text{ g} \rightarrow \frac{£1·00}{2} \rightarrow £0·50$$

$$250 \text{ g} \rightarrow \frac{£1·00}{4} \rightarrow £0·25$$

$$125 \text{ g} \rightarrow \frac{£1·00}{8} \rightarrow £0·12\frac{1}{2}$$

C   Now try these :

1. A kilogramme of chocolates costs £1·20. What is the cost of 500 g? 250 g? 125 g?

2. If corned beef costs £0·68 per kilogramme, what is the cost of 500 g? 250 g? 125 g?

3. If butter is £0·07½ for 125 g, what is the cost of a kilogramme?

4. Work out the cost of a kilogramme of cheese if 250 g costs £0·11½?

5. Work out the cost of 375 g of toffees at 56 p per kilogramme.

If we know the cost of 1 it is easy to find the cost of 10 and then 20 or 30.

1 jigsaw costs £0·37	X **10**	10 jigsaws cost £3·70
1 magazine costs £0·09		10 magazines cost £0·90
1 book costs £1·23		10 books cost £12·30

Now try these :

A   1 dictionary costs £1·78.   10 dictionaries cost £▨
         1 comic costs £0·03.   10 comics cost £▨
         1 scout's knife costs £0·32.   10 scouts' knives cost £▨

B   1 pencil costs £0·04.   20 pencils cost   £▨
         1 rubber costs £0·03½.   20 rubbers cost   £▨
         1 ball pen costs £0·17.   30 ball pens cost £▨
         1 peach costs £0·04.   50 peaches cost   £▨

1 stamp costs £0·03	X **100**	100 stamps cost £3·00
1 exercise book costs £0·13		100 books cost £13·00
1 felt tipped pen costs £1·17		100 pens cost £117·00

C   Now try these :

         If 1 l costs £0·08, 100 l cost   £▨
         If 1 kg costs £1·07, 100 kg cost £▨
         If 1 m costs £0·70, 100 m cost   £▨

D   If 1 tennis ball is worth £0·30, 200 are worth   £▨
         If 1 football is worth £4·10, 200 are worth   £▨
         If 1 hockey stick is worth £1·09, 300 are worth £▨
         If 1 plastic ball is worth £0·17, 400 are worth   £▨

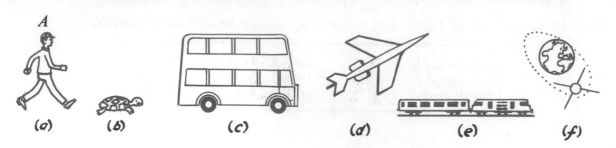

A

(a) (b) (c) (d) (e) (f)

Look at the drawings and then at these distances:

5 km, 2 500 km, 150 km, 25 000 km, 50 km, 300 m

Now write down the letter of each drawing and the distance you think would be travelled in one hour.

B

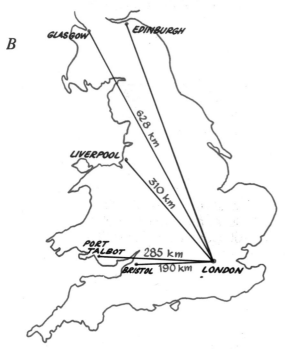

RAIL DISTANCES FROM LONDON

1. How many kilometres were travelled by a business man who made two return journeys from London to Liverpool?

2. *The Bristol Pullman* left Bristol at 13.15 and arrived in London at 15.15. What was its average speed in kilometres per hour?

3. What was the average speed in km/h of *The South Wales Pullman*, which left London at 17.40 and arrived in Port Talbot at 20.40?

4. *The Flying Scotsman* left London at 10.00 and reached Edinburgh at 16.00. Its average speed was 104·5 km/h. What is the rail distance from London to Edinburgh in kilometres?

5. A train leaves London at 21.10 and travels to Glasgow at an average speed of 62·8 km/h. At what time does it reach Glasgow?

C 1. A cyclist travelled at an average speed of 16 km per hour to a town 40 kilometres away. How long did he take?

2. A motorist travelled at an average speed of 70 km/h for $3\frac{1}{2}$ hours and then at an average speed of 80 km/h for $\frac{3}{4}$ hour. How far did he travel?

3. 100 METRES PER SECOND. What is this speed in KILOMETRES PER HOUR?

A On this drawing of a carpet we know that 1 centimetre on the drawing corresponds to 50 centimetres on the real carpet. This scale can be written :

$$1 \text{ cm} \longrightarrow 50 \text{ cm.}$$

We say : ' 1 centimetre corresponds to 50 centimetres '.

It is also true that 50 cm on the carpet corresponds to 1 cm on the drawing :

$$50 \text{ cm} \longrightarrow 1 \text{ cm.}$$

The scale can therefore be written by combining the two :

$$1 \text{ cm} \longleftrightarrow 50 \text{ cm}$$

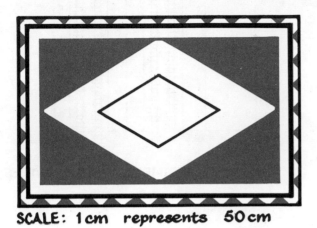

SCALE: 1cm represents 50cm

PLAN OF A CARPET

1. What is the length of the real carpet?
2. How much greater is the length than the width?
3. Calculate the perimeter of the carpet.
4. Using the same scale make a drawing of a carpet 4 m by 3 m.

B

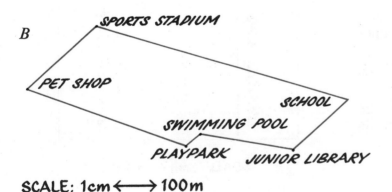

SCALE: 1cm ⟷ 100m

PLAN OF CYCLING ROUTE

1. Paul cycled from school to the pet shop, first calling at the library. How far did he cycle?
2. What distance did Paul cover when he cycled from the junior library to the sports stadium and back?

C MAP OF AN UNDERGROUND RAILWAY

NEWPARK   DALWAY   STANTON MILLINGTON   BLACKPOND   DILBANK

Scale : 1cm ⟷ 2km

1. What is the distance from Millington to Blackpond?
2. How much greater is the distance between Dalway and Blackpond than that between New Park and Stanton?

A   DRAWING OF A RAILWAY STATION AS IT LOOKS FROM AN AEROPLANE

1. What is the length of the smallest platform?

2. If a train two hundred metres long stopped at platform 5, what length of the train would be outside the platform?

3. How many metres longer is platform 3 than platform 6?

4. A passenger walked the length of platform 4 five times. How many kilometres did he walk?

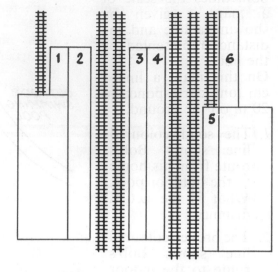

SCALE 1 CENTIMETRE ⟷ 40 METRES

B   PLAN OF A QUOIT TENNIS COURT

1. What is the distance between the two service lines on the actual court?

2. Work out the width of the court.

3. How much greater is the length of the court than its width?

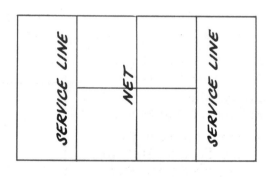

SCALE 1cm ⟷ 1m

C   PLAN OF A GARAGE

1. What is the width of the real garage?

2. How many metres are there to spare between the front of the car and the end of the garage?

3. What is the length of the car?

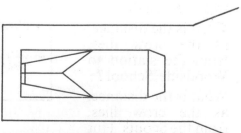

SCALE 1 CENTIMETRE ⟷ 1 METRE

A   Sometimes the scale of a map is given by showing a line and the distance represented by the line.

On this map, a line 1 cm long corresponds to 20 m on the ground.

1. The solid coloured line shows Bob's route from his home to the outdoor pool. What is the actual distance?

2. The broken coloured line shows Bob's route to the indoor baths. How much greater is the distance to the outdoor pool?

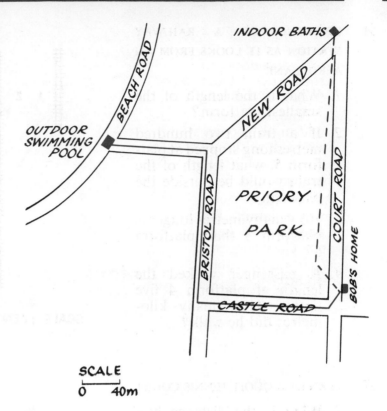

SCALE
0   40m

B   MAP OF COUNTRYSIDE AS IT LOOKS FROM AIR

Study the scale and answer these questions:

1. How far is it from the Scouts' Hut to Woodside School?

2. Andy cycled from the Scouts' Hut to Cuckoo Farm. What was the distance?

3. What is the distance, as the crow flies, from the station to Woodside School?

4. What is the distance, as the crow flies, from the Scouts' Hut to Cuckoo Farm?

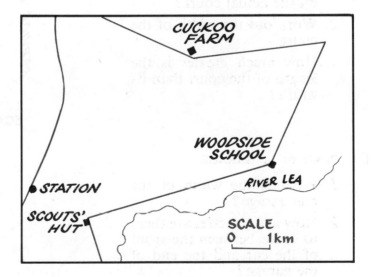

A 1. How many kilo-
metres more is it
from London to
Athens than from
London to Madrid ?

2. A plane flew from
London to Rome
and back without
refuelling. What
distance did it fly?

3. An aircraft flew
from Shannon to
London and then
on to Copenhagen.
What was the total
distance ?

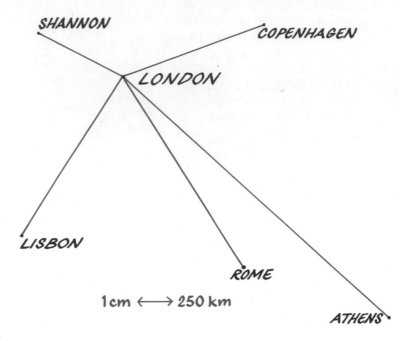

1cm ⟷ 250 km

During the war these ships
sailed from the island on a
zig-zag course to avoid
enemy submarines.
Work out the distance
covered by each ship be-
fore it set sail on a course
due north.

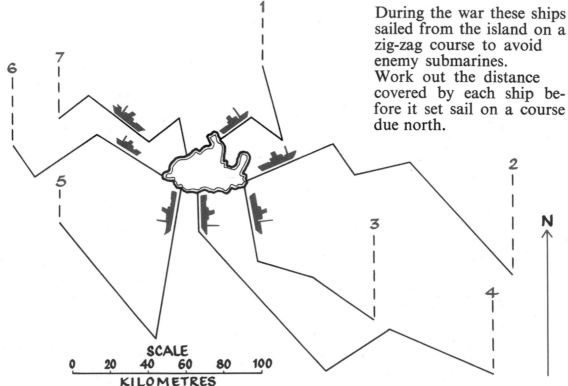

SCALE

0   20   40   60   80   100

KILOMETRES

Before the Chinese had a written number system, they made numerals by laying out bamboo rods on a counting board.

The earliest numerals were shaped like this:

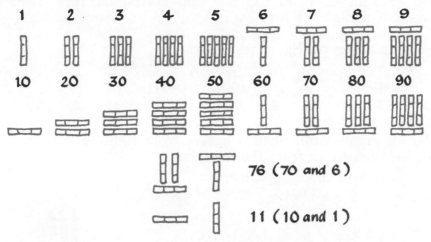

A   Now cut up some milk straws, or use matchsticks, for rods of your own.
    Use your rods to show these numbers. Make drawings of your rod numerals in your book.

$$35 \quad 47 \quad 66 \quad 19 \quad 50$$

B   To represent hundreds the unit symbols were used again, like this:

181

This is similar to our system, when the same numeral can represent nine (9) or nine hundred (900).

To represent thousands the tens symbols were used again, like this:

2 323

The *position* of each symbol had to be thought about carefully.

Write these numbers in Arabic numerals:

C   Now set out your rods to represent these numbers :

$$433 \quad 2321 \quad 7563 \quad 990$$

Record your answer by making drawings of the rods.

It is believed that the earliest written numbers were used in Egypt five thousand years ago. The Egyptians wrote on a rough kind of paper called 'papyrus' which was made out of water reeds gathered from the banks of the River Nile. The papyrus was cut into widths of about 25 centimetres, and then rolled into scrolls. The Egyptians wrote with a brush dipped in an ink made of black powder and water.

These are the numerals the Egyptians used :

Usually the Egyptians wrote their numbers from right to left. Sometimes they wrote from left to right, or from the top down.

For example, 3911 could be written:

A   Write the numbers below from left to right using Egyptian numerals. First, here are two examples:

378	4125	3074	916
17	5001	203	1270

B   Look at these numbers carefully to see if they are written from left to right, from right to left, or downwards. Write the numbers in Arabic numerals.

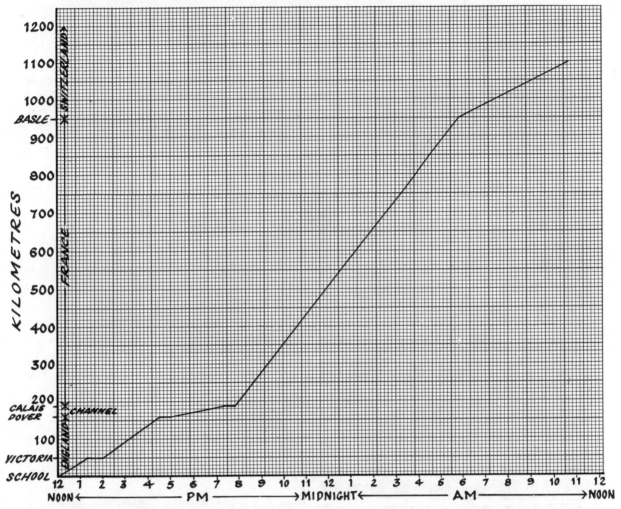

A   This is a journey made by the pupils of Danesfield County Primary School, Marlow.

1. How long was spent altogether waiting at Victoria, Dover and Calais?

2. How long did the Channel crossing take?

3. What was the average speed of the train between Calais and Basle?

4. Work out the distance travelled by train across Switzerland.

5. Look at the slope of the graph. How can you tell the part of the journey which was covered at the highest speed?

B   Write out a time-table of any school journey you have made. Find the distances between places, and then illustrate the journey by drawing a graph.

A   KNOTTED CORDS

The Zuñi Indians of America had a clever system of knot numerals. A large knot represented ten, a medium knot five and a small knot one.

A small knot before a medium knot represented four (5−1). A small knot after a medium knot represented six (5+1). In the same way a small knot before a large knot represented nine (10−1) and a small knot after a large knot, eleven (10+1).

Write down the numbers represented by these drawings:

B   TALLY STICKS

The tally stick was a length of wood on which notches were cut to show numbers and amounts of money. Any number could be shown by the position and size of the notches, and this was the usual way of keeping accounts during the Middle Ages.

When the tally was finished it was split lengthwise so that there were two parts exactly alike. One half was kept by the buyer and the other half by the seller. If there was an argument about the price the two parts could be put together to see if they matched or 'tallied'.

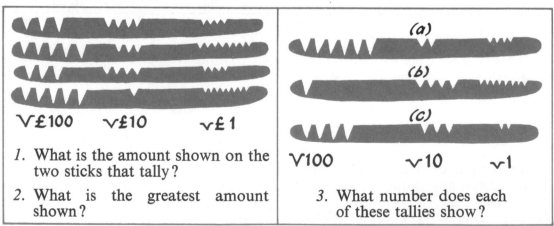

1. What is the amount shown on the two sticks that tally?

2. What is the greatest amount shown?

3. What number does each of these tallies show?

C   Try to find some soft wood and cut a tally to represent the number two hundred and thirty-four.

Here are two cardboard strips, one lying on top of the other.

When the coloured strip is turned away from the white strip, or the white strip is turned away from the coloured strip, an angle is formed.

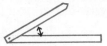

When the strips are turned farther away from each other, the angle is greater.

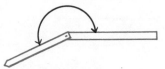

If the strips are turned more, the angle is greater still.

When the coloured strip or the white strip has been turned completely and brought back to its starting position, it has made ONE REVOLUTION.

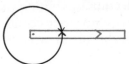

Compare the following pairs of angles. You will need a small sheet of very thin paper. Trace the arms or sides of one of each pair, and then fit the tracing over the other angle. Which is the greater angle in each pair?

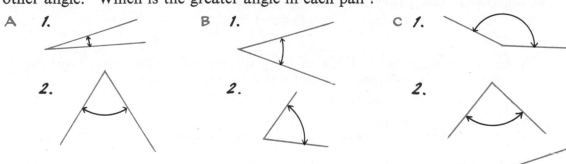

A   *1.*   B   *1.*   c *1.*

    2.   *2.*   *2.*

Does the length of the sides or arms of the angle make any difference to the size of the angle?
You can make a very simple piece of apparatus for comparing angles.
Join two strips of paper about 6 cm by 1 cm, with a paper clip.

A   To measure the size of angle we could make units of measurement of our own.

If we imagined this angle on a clock face we could say that the angle measured 3 units. In that case, an hour space would be the unit for measuring the amount of turning.

Estimate the size of these angles, using the hour space as a unit of measurement.

1.   *2.*   *3.*   *4.*

 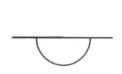  

This, of course, is not a very accurate way of measuring angles. Just as we have standard measures for length, weight, capacity, time and area, we have a standard unit for angle measurement.

B   The standard unit for measuring angles is the DEGREE.

When the hand of a clock makes one complete revolution it moves through 360 degrees, or 360°. The symbol ° means ' degree ' or ' degrees '.

1. How many degrees are there in (*a*) a quarter of a revolution? (*b*) three-quarters of a revolution? (*c*) half a revolution?

2. How many degrees are there in a right angle?

3. How many degrees are there in a straight angle?

4. How many degrees are there in half a right angle?

Our method of measuring angles was first used by the Babylonians. It had been estimated that the earth took 360 days to complete its orbit around the sun.

The Babylonians, therefore, decided to make the year a period of 360 days so that it would complete a circle of the seasons. Each day would be one part, or one *degree*, of that circle.

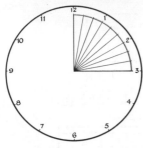

1. Here is a clock face with a part of it marked off in degrees.

Through how many degrees does the hand of a clock turn:

(*a*) from 12 to 1?   (*b*) from 12 to 6?
(*c*) from 2 to 3?   (*d*) from 1 to 3?

2. Through what angle does the minute hand of a clock move in (*a*) ¾ hour, (*b*) 20 minutes, (*c*) 1 hour?

3. Here is a simple compass card.

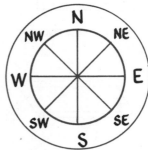

Through how many degrees does a person turn when he turns from:

(*a*) N through E and S to W? (*b*) N through E to SE?
(*c*) E through NE to N? (*d*) W through S and E to NE?

4. Angles and degrees are used a great deal in gunnery.

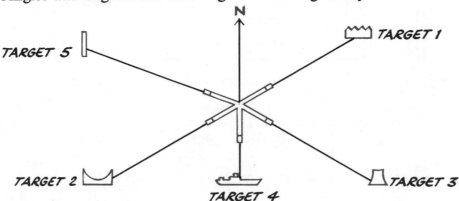

Look at the drawing above and estimate through how many degrees the gun turned in moving:

    (*a*) from target 1 to target 2   (*c*) from target 3 to target 4
    (*b*) from target 2 to target 3   (*d*) from target 4 to target 5

5. Calculate the size of these angles:

   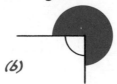   

You will need cardboard strips and paper fasteners,
or Meccano strips and bolts.

Use your cardboard strips or Meccano strips to make rectangles like these:

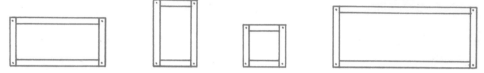

Press down on these rectangles and you will see that they can be pushed out of shape:

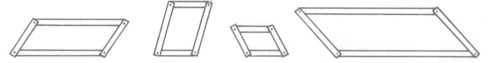

Now fasten strips across each shape, like this:

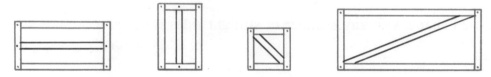

Press down on each to see if it can be moved into a different shape.

You will find that two of the frames are fixed in shape even if you pull them or press on them. We say that these frames are RIGID. Each has been made rigid by adding a strip to make the frame into two *triangles*.

The triangle is a very strong shape and is used a great deal by the engineer, especially when he constructs girder bridges, piers and towers.

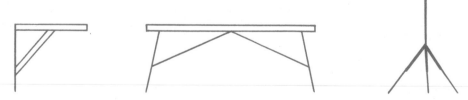

Write a list, or make drawings, of objects in which the triangle has been used because of its great strength.

A   Which of these are rigid frames?

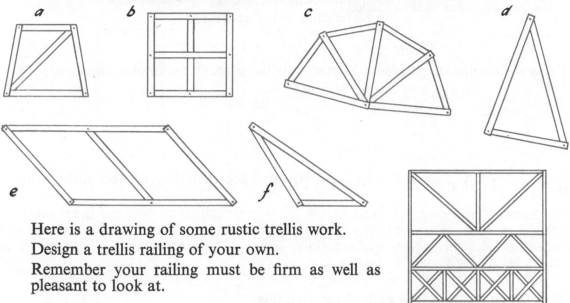

Here is a drawing of some rustic trellis work.
Design a trellis railing of your own.
Remember your railing must be firm as well as
pleasant to look at.

B   A closed shape with any number of straight sides is called a POLYGON.

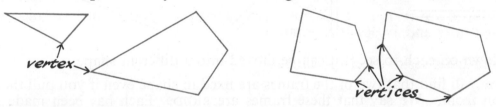

Each of these points is called
a VERTEX.

The plural of VERTEX is VERTICES.

If all the sides of a polygon are equal, it is called a REGULAR POLYGON.

Use your strips to make shapes similar to these regular polygons. How
many cross strips are needed to make each frame rigid? You will find it
much easier to work from one vertex.

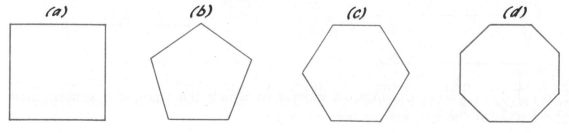

Any shape which has four straight sides is called a QUADRILATERAL. *Quad* means ' four ' and *lateral* means ' sided '.

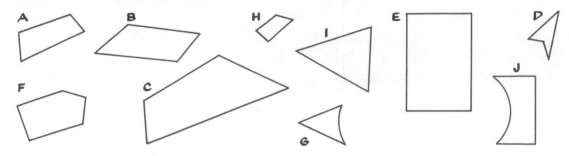

Which of the shapes above belong to the set of quadrilaterals?

    Look at the quadrilaterals below. They are all alike because they have the same number of sides, but they are also alike in other ways. Measure the sides of each quadrilateral. Now make a right angle by folding a sheet of paper. Test the angles of each quadrilateral.

These shapes are *rectangles*. Write a sentence to say what you have discovered about the sides and angles of a rectangle.

Here we have a set of special rectangles:

Measure the sides of each to see why it is special. What name do we give to these shapes?

Make a rectangle with card-board strips, like this:

Press down on one corner of the rect-angle to make this shape:

These shapes make another special set of quadrilaterals. They are called PARALLELOGRAMS. We can see that the lengths of the sides have not been altered, but the sizes of the angles have been changed.

A   Copy this table and in place of each □ put a tick or a cross.

SHAPE	Always has 4 sides	Always has 4 right angles	Opposite sides always parallel	Opposite sides always equal
A RECTANGLE	□	□	□	□
A PARALLELOGRAM	□	□	□	□

Make a square with your strips and paper fasteners.

Press down on the square until it is shaped like this:

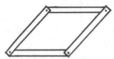

We can see that the sides are still equal and parallel. This shape is sometimes called a 'diamond'. In mathematics it is called a RHOMBUS.

Copy this table and in place of each □ put a tick or a cross.

SHAPE	Always has 4 equal sides	Opposite sides always parallel	Always has 4 right angles
A SQUARE	□	□	□
A RHOMBUS	□	□	□

B   There is one more member of the quadrilateral family. It is called a TRAPEZIUM. Can you see what is special about trapeziums?

These trapeziums will probably give you the clue:

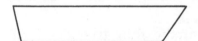 

C   Copy the lists below and do some 'mapping'. The first one is done for you.

SQUARE	Quadrilateral Rhombus Parallelogram Rectangle Trapezium	RECTANGLE	Quadrilateral Rhombus Parallelogram Trapezium Square
RHOMBUS	Quadrilateral Parallelogram Rectangle Trapezium Square	TRAPEZIUM	Quadrilateral Rhombus Parallelogram Square Rectangle

| You will need thin card and scissors. |

A   The line joining the opposite corners of a rectangle is a *diagonal*.

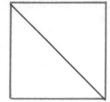 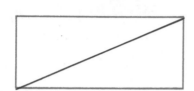 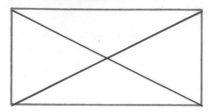

On a sheet of thin card draw a square with sides of 5 cm. Now draw the two diagonals and cut out the four triangles.

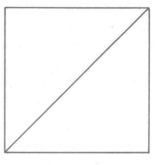

1. Place two of the triangles together to make a small square, as in the diagram.

2. Now place a triangle against this shape to make it into a trapezium. Copy the trapezium, or place the pieces on a page and draw around the shape.

3. Take the other triangle and add it to the trapezium to make a large triangle.

4. Remove one small triangle and place it so that the four pieces make a parallelogram.

5. Remove one small triangle from the parallelogram and place it so that the four pieces make a rectangle.

6. Remove one piece and place it so that the four triangles form another trapezium.

B   There are four ways of making this *parallelogram* into a *trapezium* by adding a *triangle*. See if you can discover the four ways.

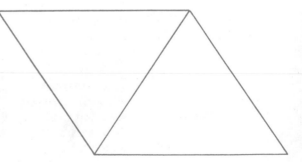

We have already learned that we can find the *surface measurement*, or *area*, by covering that surface with objects of the same shape and size. Any objects which pack together can be used.

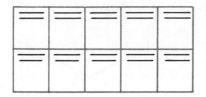

The surface of this table can be covered exactly by 10 exercise books.

The surface of this strip can be covered by 10 triangular stamps.

We have also learned that the best unit to use for surface measurement is the square. Can you think why?

 The area of this shape is 12 units.

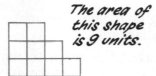 The area of this shape is 9 units.

The area of this shape is 31 units.

Find the number of square units in each of the shapes below. This is the square unit used to measure the area of each shape: ☐

(a)

(b)

(c)

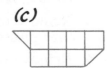

Use the same square unit and draw three shapes, each with an area of 12 square units.

Do you remember the quick way of calculating the area of rectangles?

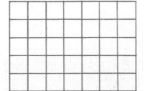

There are 7 square units in a row and there are 5 rows.
$(7 \times 5) = 35$.
Area of rectangle is 35 square units.

Calculate the area of these rectangles in square units.

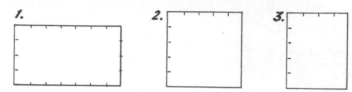

If we all chose our own square units it would be impossible to compare the sizes of surfaces. So we have standard units of area which are related to the standard units of length.

The unit used for measuring small areas is the *square centimetre* (cm²).

Find centimetres on your ruler. Now measure the length and breadth of rectangles A, B, C, D and E below. Calculate each area in square centimetres like this :

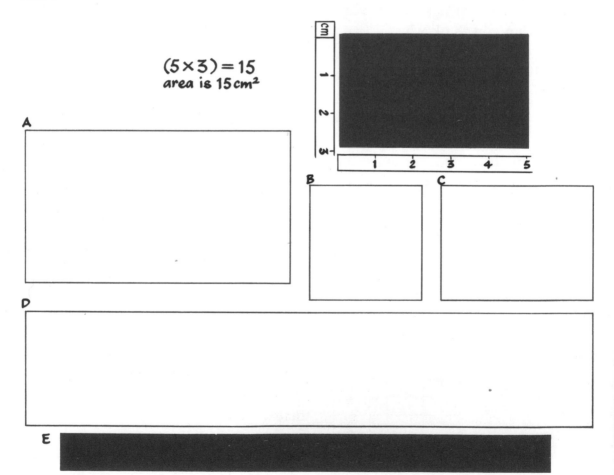

$(5 \times 3) = 15$
area is 15 cm²

Sometimes we have to find the area of an irregular shape. It may be possible to divide the shape into rectangles and find their total area, like this:

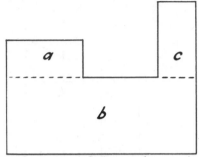

area of rectangle *a*:   2 sq cm
area of rectangle *b*: 10 sq cm
area of rectangle *c*:   2 sq cm
total area: 14 sq cm

Now calculate the area of these shapes in square centimetres:

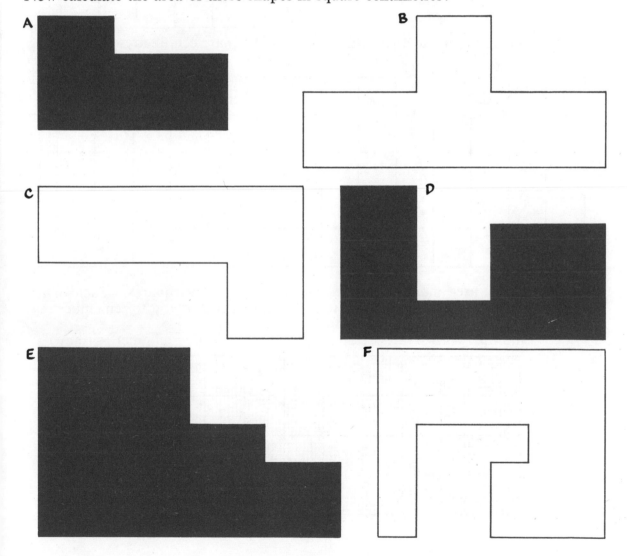

Many shapes, of course, cannot be divided into rectangles. We cannot calculate their areas by measuring the length and breadth.

This leaf has been drawn on centimetre squares. If we count up the number of square centimetres the leaf covers we shall know its *approximate* area. When half the square or more is covered we count it as a whole square. If less than half the square is covered it is not counted at all.

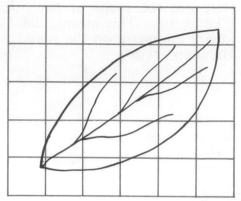

A   Find the approximate area of these shapes in square centimetres (cm²) :

1.   *2.*   *3.*

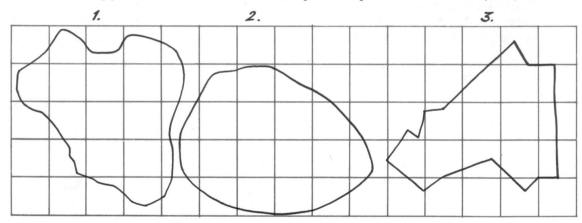

B   To measure area more accurately, we can use smaller squares. Each of the small squares below is a millimetre square. It will help you to remember that each larger square contains 100 millimetre squares.
   Find the approximate area of these shapes in square millimetres (mm²).

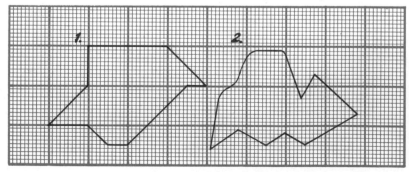

You will need a sheet of tracing paper.

A   Here are the maps of two islands drawn to a scale of 1 centimetre to 1 kilometre. This means that 1 square centimetre on the map represents 1 square kilometre on the island. Estimate the area of each island in square kilometres (km²).

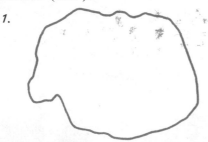

1.

2.

Draw 1-centimetre squares on a small sheet of tracing paper. Lay the tracing paper over each island in turn and count the number of square centimetres needed to cover each one. What is the area of each island in square kilometres?

B   Below are the plans of two playgrounds. Both plans are drawn to a scale of 1 centimetre to 10 metres.

1. How many square metres (m²) are represented by one square centimetre?
2. Estimate the area of each playground in square metres (m²).
  Now use your tracing paper to find the area of each playground.

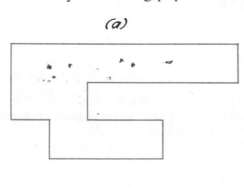

(a)

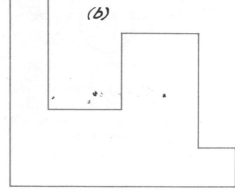

(b)

C   The shapes below are drawn to a scale of 1 cm to 1 m. Calculate the area of each in m².

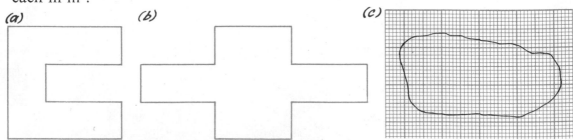

(a)   *(b)*   *(c)*

A

DAYLIGHT	DARKNESS

This diagram represents the hours of daylight and darkness on a certain day.

1. How many hours of daylight were there?

2. What scale was used? Say what length represents one hour.

3. Do you think the day was in mid-summer, spring or mid-winter?

B

MARCH   JUNE   SEPTEMBER

These diagrams show the progress of a school's swimming-pool fund. The target was £600.

1. What percentage of the target had been saved by March?

2. How much was saved between June and September?

3. By September, how much more was needed to reach the target?

C

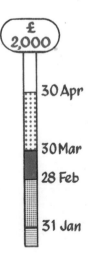

The diagram on the left shows the amount of money collected for a church fund.

1. How much money was collected during February?

2. What was the largest amount collected in a month?

3. How much more is needed to reach the target?

4. What amount is represented by the diagram on the right?

A   This rectangle represents the favourite dogs of 330 pupils. By finding what part of the total each section represents, you will be able to answer the questions below.

THE FAVOURITE DOGS OF 330 PUPILS AT LAKESIDE SCHOOL

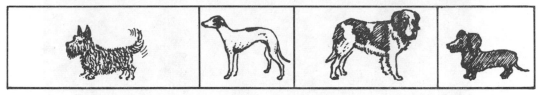

1. How many pupils voted for the Scots terrier?

2. How many pupils altogether voted for the greyhound and Scots terrier?

3. How many more votes had the St. Bernard than the dachshund?

B

THE FAVOURITE GAMES OF MEMBERS OF A YOUTH CLUB

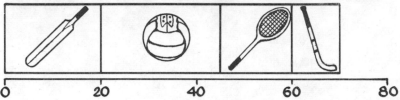

1. How many members like football best?

2. How many more members voted for cricket than for tennis?

3. How many more members voted for tennis than for hockey?

4. Do you think there were more boys or more girls in the club? Give your reason.

C

HOW BOB SPENDS HIS WEEKLY POCKET MONEY

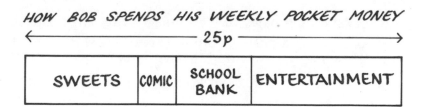

1. How much does Bob spend altogether on sweets and his comic?

2. How much does Bob save each week?

3. How much does Bob have left for entertainment after spending a 5p piece to go to Saturday morning cinema?

4. What is the price of Bob's comic?

A

BELGIAN COASTAL RESORTS
8 Days by Rail/Boat

Tour No.	Centre	Under 14	14–21	Adults
B1	OSTEND			
	Easter	£14·20	£16·10	£19·05
	Whitsun	£15·25	£17·15	£20·30
	Summer	£15·25	£17·15	£20·30
B2	BLANKENBERGE			
	Easter	£14·35	£16·30	£19·20
	Whitsun	£15·40	£17·35	£20·70
	Summer	£15·40	£17·35	£20·70
B3	WENDUINE			
	Easter	£13·90	£15·80	£18·70
	Whitsun	£14·25	£16·15	£19·30
	Summer	£14·25	£16·15	£19·30
B4	BRUGES			
	Easter	£14·35	£16·30	£19·20
	Whitsun	£15·40	£17·35	£20·70
	Summer	£15·95	£17·85	£21·40

1. Work out the cost of sending a party of 10 junior school pupils under 14 years to Bruges at Whitsun.
2. How much more would the Bruges holiday cost than a holiday in Wenduine during Easter?
3. Find the cost of sending a party of 7 adults to Bruges at Easter.
4. The total cost of sending a party of pupils (ages 14–21) to Ostend at Easter was £177·10. How many pupils were in the party?

Now make up two or three examples of your own for your friends to solve.

B   1. The total cost of sending a party of 3 adults and 10 pupils to Ostend at Christmas was £212·70. The cost for each adult was £18·90. What was the cost for each child?

      2. The cost of sending a party of eleven pupils to Interlaken was £362·45. In addition, each pupil paid £1·87½ for an excursion. What was the total cost for each pupil?

Here is a simple calculator which can be used to save time in working out costs.

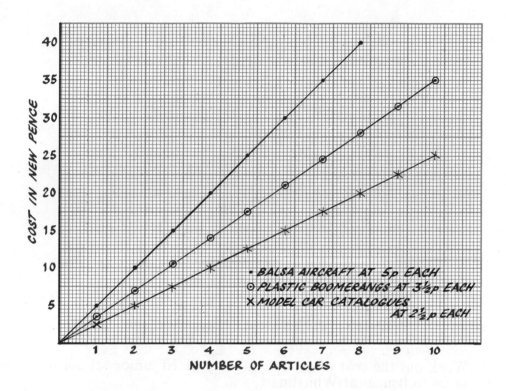

Use the graph to answer these questions :

1. What is the cost of 7 balsa aircraft?
2. Find the cost of 5 plastic boomerangs.
3. A boy needs another 10½p to buy 8 plastic boomerangs. How much has he?
4. Work out the total cost of 7 balsa aircraft, 7 plastic boomerangs and 7 catalogues.
5. How many boomerangs can be bought with 24½p?
6. How much more must be paid for 5 aircraft than for 5 boomerangs?
7. A boy had 2½p change from a 50p piece after buying 3 catalogues and some balsa aircraft. How many aircraft did he buy?

Now use the graph to make up some examples of your own.

We have already learned that the distance around a shape is called its *perimeter*.

Find the perimeter of the top of your desk in centimetres.

Find the perimeter of your classroom in metres.

A   PERIMETER OF POLYGONS

A polygon is a closed shape with any number of straight sides.

Find the perimeter of these polygons.

1.

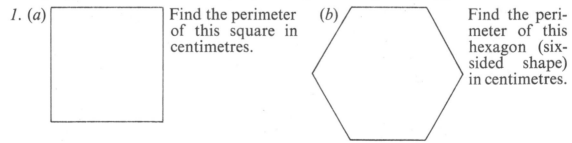

Line AB is ☐ cm.
Line BC is ☐ cm.
Line AC is ☐ cm.

The perimeter of this triangle is ☐ cm.

2.

Line MN is ☐ cm.
Line NO is ☐ cm.
Line OP is ☐ cm.
Line MP is ☐ cm.

The perimeter of this quadrilateral is ☐ cm.

Now write this sentence in full:

The perimeter of a polygon is the ☐ of the lengths of its sides.

B   PERIMETER OF REGULAR POLYGONS

A polygon with all its sides of equal length is a *regular* polygon.

1. (a)   Find the perimeter of this square in centimetres.   *(b)*   Find the perimeter of this hexagon (six-sided shape) in centimetres.

2. What do you notice about the angles of a regular polygon?

3. Find the perimeter of a regular pentagon (five-sided shape) with sides of 7·5 cm.

4. The perimeter of a regular octagon (eight-sided shape) is 28 cm. What is the length of a side?

Use a length of thread to find the perimeter of each of the regions (*a*) and (*b*) in centimetres.

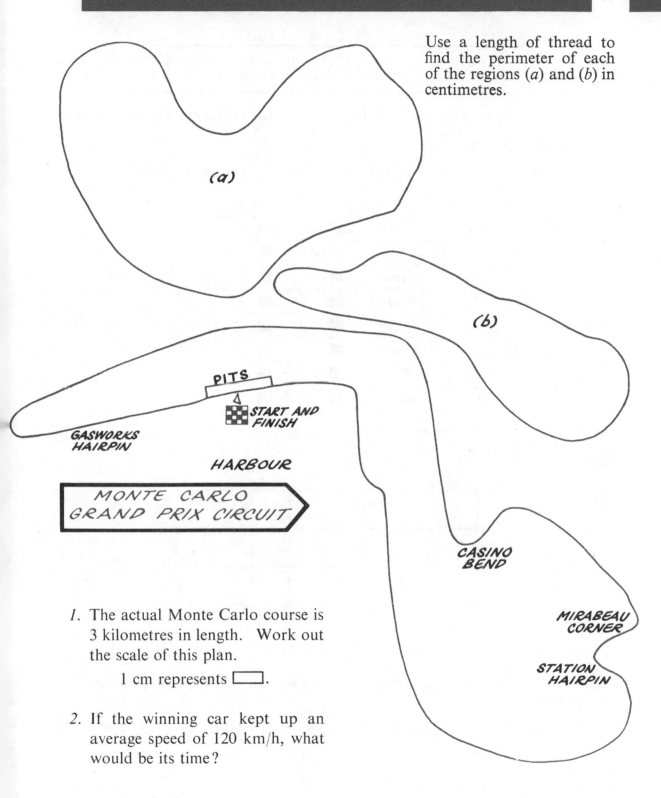

1. The actual Monte Carlo course is 3 kilometres in length. Work out the scale of this plan.

   1 cm represents ⬜.

2. If the winning car kept up an average speed of 120 km/h, what would be its time?

> You will need 24 separate centimetre squares of cardboard and a sheet of 1 centimetre squared paper.

Do you think that rectangles with the same area also have the same perimeter?

Copy this table :

RECTANGLES WITH AN AREA OF 24 SQUARE CENTIMETRES (24 cm²)

Width	Length	Perimeter
1 cm	24 cm	50 cm
2 cm	12 cm	■
3 cm	■	■
4 cm	■	■
6 cm	■	■
8 cm	■	■
■	■	■
■	■	■

Using all your cut-out squares each time, make rectangles with the widths shown in the table. Your 3-cm-wide rectangle will look like this :

Make a drawing of your rectangles on the squared paper and then fill in the blanks in the table.

You must work out the last two widths for yourself.

Make a drawing of your rectangles on the squared paper and then fill in the blanks in the table.

Can you think what length the rectangle would be if the width were ½ cm?

Do all rectangles which have the same area have the same perimeter?

You will need 24 centimetres of string, with which to make rectangles of different shapes, and a sheet of 1 centimetre squared paper.

Do you think that rectangles which have the same perimeter also have the same area?

Starting with a width of 1 cm, write down the length and area of all the rectangles you could make with your 24-cm length of string. Record your results in a table, like this :

RECTANGLES WITH A PERIMETER OF 24 CM

Width	Length	Area
1 cm	11 cm	11 cm²
2 cm	10 cm	■
3 cm	■	■
4 cm	■	■
5 cm	■	■
6 cm	■	■
7 cm	■	■
8 cm	■	■
■	■	■
■	■	■
■	■	■

Draw these rectangles on your squared paper. Finish off the table. What special rectangle gives the greatest area?

What would be the length of a rectangle with a width of $\frac{1}{2}$ cm? What would be the area of this rectangle ?

Do all rectangles which have the same perimeter have the same area?

Can you think of a way of showing your results on a graph?

A   MAGIC CIRCLES

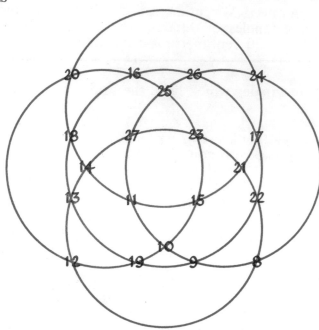

Add the eight numbers on each circle. Is the sum of these numbers the same on each of the five circles? If so, these are *magic circles*.

B   *1.* What is half the difference between 478 and 4780?

2. Work out the number that is equal to 37 times the sum of 364, 463, 346, 436, 634 and 643.

3. Calculate the number which must be added to the product of 17 and 19 to make 2230.

4. What number is 407 more than the product of 407 and 9?

5. Calculate the number that is twelve times as great as the difference between 4786 and 6847.

6. What number is equal to twice the product of 45 and 67?

7. Give the number that is 29 less than the quotient of 1406 and 37.

8. What must be added to the sum of 4786 and 2314 to equal the product of 47 and 230?

C   To solve these equations, we must find the number to replace **n**.

1. $(96+8)-12 = \mathbf{n}$ *2.* $(65-7)+11 = \mathbf{n}$

3. $(8 \times 9) \div 12 = \mathbf{n}$ *4.* $(7 \times 6) \div \mathbf{n} = 7$

5. $(\mathbf{n} \times 6) \div 4 = 12$ *6.* $(72 \div \mathbf{n}) \div 3 = 3$

FINDING THE AMOUNT OF SPACE INSIDE A CONTAINER

> You will need some containers of different sizes, a box of cubes, water and rice or sand.

Use a teacup, or if the containers are small an egg-cup, to fill each container with rice, water or dry sand. In this way you can find out which container has the largest VOLUME (that is, which holds most).

Now arrange them in order of size.

Find some boxes of different shapes.

Use a small box like a matchbox or a nib box to fill each container with sand. How many boxfuls of sand does each hold?

Arrange the boxes in order of size.

Now use cubes to measure the amount of space in the boxes.

Use cubes to measure the amount of space in the jam jar. What do you notice when you try to pack cubes into a jar?

We know that not all matchboxes or nib boxes are the same size. Just as we have standard units in length, so we must have standard units to measure *volume*.

Take a ball of plasticine and make a cube with sides of 1 centimetre. This is a CUBIC CENTIMETRE (cm^3).

Now make this cubic centimetre into different shapes, like this :

> You will need a piece of thin card.

Now make a container for measuring volume in cubic centimetres (cm³).

Cut out a square 6 cm by 6 cm. Mark off the sides in 2 cm, and join up the dots.

Draw this shape. Mark in the dotted lines, which are 'fold' lines. Remember to cut round the flaps. Use Sellotape or glue to stick the flaps.

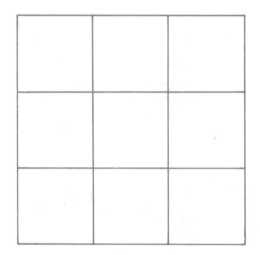

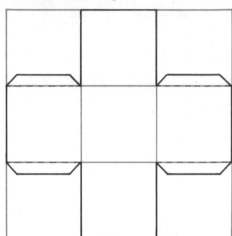

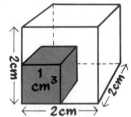

This container holds 8 cm³.
How many times must you fill the cube with sand to fill a cup?
What is the volume of the cup in cm³?

Would you use solid cubes or the cubic inch measure to find the volume of:

    (*a*) a shoe box?   (*b*) a basin?   (*c*) a bottle?   (*d*) a chalk box?

The cubes below make a shape just like the box.

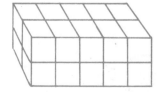 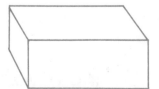

The box takes up as much space as 20 *unit cubes*.
The volume of the box is 20 *cubic units*.

A   Find the volume (the number of cubic units) in each of these arrangements:

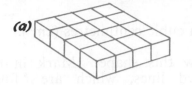 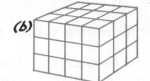 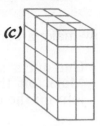

B   Can you calculate the volumes of these solid shapes?

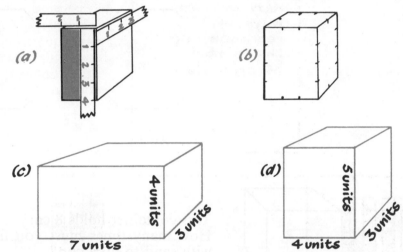

C   Calculate the number of cubic centimetres in this solid. One dimension (measurement) has been given you.

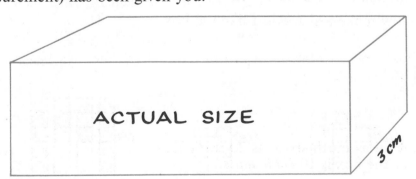

ACTUAL SIZE

3 cm

1. Find the difference between 9783×103 and 9793×103 in your head.

2. Is this statement true or false: $\frac{1}{6} + \frac{1}{4} = \frac{1}{10}$?

3. Find the numerator or denominator which can replace $\square$ in each of these statements.

$$1\tfrac{3}{7} = \frac{\square}{7} \qquad\qquad \tfrac{1}{7} = \frac{\square}{14}$$

$$\tfrac{3}{4} = \frac{\square}{24} \qquad\qquad \tfrac{1}{5} = \frac{3}{\square}$$

4. Write the smallest fraction possible with the digits 6, 7, 8 and 9. Use one digit for the numerator and three digits for the denominator.

5.

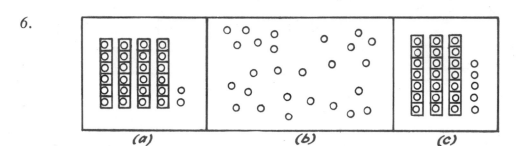

Here are some ways of representing the number of triangles.

VI   11 _five_
ᑎᕼᒍ I   4 + 2
3 × 2   20 _three_

We know that there are many ways of representing a number. Can you represent the number of stars shown in at least ten ways? Study the example in the rectangle.

6.

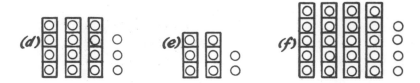

(a)   (b)   (c)

The example above shows the same beads (*a*) collected in trays each holding six, (*b*) tipped out, and (*c*) collected in trays each holding seven.

We can represent the same number of beads like this: 42_{six} or 35_{seven}.

Write the number of beads shown in each of the pictures below in base six, base seven and base eight.

(d)   *(e)*   *(f)*

A

> Each of the letters of the alphabet is written on a small slip of paper which is then folded and placed in a box. The slips are mixed thoroughly and you are then allowed to draw out one slip.

1. What is the chance of drawing out the first letter of the alphabet?
2. What is the chance of drawing out one of the two last letters of the alphabet?
3. What is the chance of drawing out a vowel?
4. What is the chance of drawing out one of the letters of the name Sam Smith?

B

$$P = \{prime\ numbers\ less\ than\ 20\}$$
$$S = \{square\ numbers\ less\ than\ 20\}$$
$$T = \{triangular\ numbers\ less\ than\ 20\}$$

Write the letter P, S or T to show to which set the following numbers belong:

(*a*) 7   (*b*) 10   (*c*) 9   (*d*) 4   (*e*) 13   (*f*) 15

C   Write down the name of the units you would use in measuring:

(*a*) the temperature of a room; (*b*) the weight of a letter; (*c*) the area of a playground; (*d*) the time it takes to run across a playground; (*e*) the distance between towns.

D   Work out the following additions, using Roman numerals:

1. DCLXXXVII   *2.* CCCLXXVIII
    CLXXXVI   CCCLXXV

E   Work out the area of a square whose perimeter is 100 centimetres.

A  Here is a clock face which can be used for 24-hour times.
Copy this table and then use the clock face to help you to fill in the missing times.

17·00	20·30	21·00		23·00			16·45
5 p.m.			3·30 p.m.		1 p.m.	8 p.m.	

B   Use the 4 clock to help you to complete this addition square:

ON 4 CLOCK

+	0	1	2	3
0	0			
1		2		
2			0	
3				2

C   This is how Carol recorded the favourite school desserts of the other girls in her class.

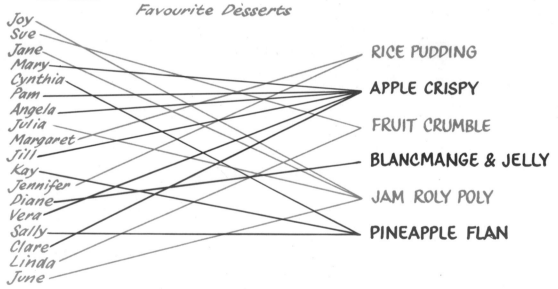

Favourite Desserts

Use the information above to complete this table.

FAVOURITE DESSERT	RICE PUDDING	APPLE CRISPY	FRUIT CRUMBLE	BLANCMANGE & JELLY	JAM ROLY POLY	PINEAPPLE FLAN
NUMBER OF GIRLS						

Find out the favourite desserts of the boys in your class. Record the information as Carol recorded hers, then make a bar chart to show this information.